# Modelling and Simulation of Fast-Moving Ad-Hoc Networks (FANETs and VANETs)

T.S. Pradeep Kumar
*Vellore Institute of Technology, India*

M. Alamelu
*Kumaraguru College of Technology, India*

A volume in the Advances in Wireless Technologies and Telecommunication (AWTT) Book Series

Published in the United States of America by
IGI Global
Information Science Reference (an imprint of IGI Global)
701 E. Chocolate Avenue
Hershey PA, USA 17033
Tel: 717-533-8845
Fax: 717-533-8661
E-mail: cust@igi-global.com
Web site: http://www.igi-global.com



Library of Congress Cataloging-in-Publication Data

Names: Kumar, T. S. Pradeep, 1976- editor. | Alamelu, M., 1981- editor.
Title: Modelling and simulation of Fast moving Ad-hoc Networks (FANETs and VANETs) / T.S. Pradeep Kumar and M. Alamelu, editors.
Description: Hershey, PA : Engineering Science Reference, an imprint of IGI Global, [2022] | Includes bibliographical references and index. | Summary: "This book enhances the modelling and simulation aspects of FANETS and VANETS and presents understanding of the protocols in mac layer and network layers for fast-moving ad-hoc networks"-- Provided by publisher.
Identifiers: LCCN 2022012080 (print) | LCCN 2022012081 (ebook) | ISBN 9781668436103 (h/c) | ISBN 9781668436110 (s/c) | ISBN 9781668436127 (ebook)
Subjects: LCSH: Ad hoc networks (Computer networks)
Classification: LCC TK5105.77 .M64 2022 (print) | LCC TK5105.77 (ebook) | DDC 004.6--dc23/eng/20220504
LC record available at https://lccn.loc.gov/2022012080
LC ebook record available at https://lccn.loc.gov/2022012081

This book is published in the IGI Global book series Advances in Wireless Technologies and Telecommunication (AWTT) (ISSN: 2327-3305; eISSN: 2327-3313)

British Cataloguing in Publication Data
A Cataloguing in Publication record for this book is available from the British Library.

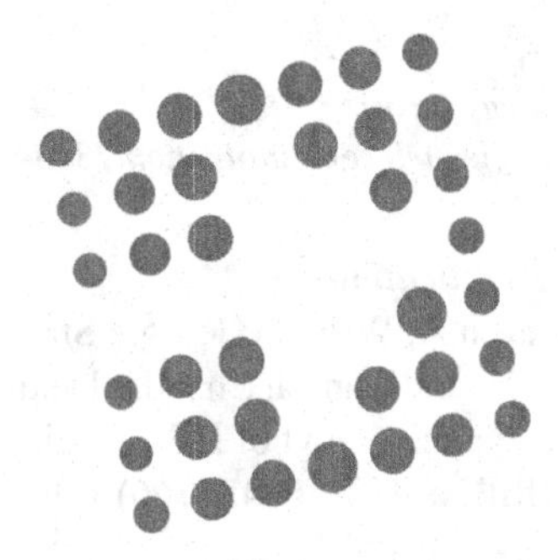

# Advances in Wireless Technologies and Telecommunication (AWTT) Book Series

ISSN:2327-3305
EISSN:2327-3313

Editor-in-Chief: Xiaoge Xu, University of Nottingham Ningbo China, China

## MISSION

The wireless computing industry is constantly evolving, redesigning the ways in which individuals share information. Wireless technology and telecommunication remain one of the most important technologies in business organizations. The utilization of these technologies has enhanced business efficiency by enabling dynamic resources in all aspects of society.

The **Advances in Wireless Technologies and Telecommunication Book Series** aims to provide researchers and academic communities with quality research on the concepts and developments in the wireless technology fields. Developers, engineers, students, research strategists, and IT managers will find this series useful to gain insight into next generation wireless technologies and telecommunication.

## COVERAGE

- Mobile Technology
- Mobile Web Services
- Wireless Sensor Networks
- Radio Communication
- Network Management
- Virtual Network Operations
- Wireless Technologies
- Mobile Communications
- Telecommunications
- Global Telecommunications

IGI Global is currently accepting manuscripts for publication within this series. To submit a proposal for a volume in this series, please contact our Acquisition Editors at Acquisitions@igi-global.com or visit: http://www.igi-global.com/publish/.

The Advances in Wireless Technologies and Telecommunication (AWTT) Book Series (ISSN 2327-3305) is published by IGI Global, 701 E. Chocolate Avenue, Hershey, PA 17033-1240, USA, www.igi-global.com. This series is composed of titles available for purchase individually; each title is edited to be contextually exclusive from any other title within the series. For pricing and ordering information please visit http://www.igi-global.com/book-series/advances-wireless-technologies-telecommunication/73684. Postmaster: Send all address changes to above address. 

## Titles in this Series

***For a list of additional titles in this series, please visit:***
***http://www.igi-global.com/book-series/advances-wireless-technologies-telecommunication/73684***

***Role of 6G Wireless Networks in AI and Blockchain-Based Aplications***
Malaya Dutta Borah (NIT Silchar, Cachar, Assam, India) Steven A. Wright (Georgia State University, USA) Pushpa Singh (Delhi Technical Campus, India) and Ganesh Chandra Deka (RDSDE,DGT, Ministry of Skill Development and Entrepreneurship, Govt of India, India)
Information Science Reference • © 2023 • 320pp • H/C (ISBN: 9781668453766) • US $250.00

***Challenges and Risks Involved in Deploying 6G and NextGen Networks***
A.M. Viswa Bharathy (GITAM University, Bengaluru, India) and Basim Alhadidi (Al-Balqa Applied University, Jordan)
Information Science Reference • © 2022 • 258pp • H/C (ISBN: 9781668438046) • US $250.00

***Achieving Full Realization and Mitigating the Challenges of the Internet of Things***
Marcel Ohanga Odhiambo (Mangosuthu University of Technology, South Africa) and Weston Mwashita (Vaal University of Technology, South Africa)
Engineering Science Reference • © 2022 • 263pp • H/C (ISBN: 9781799893127) • US $240.00

***Handbook of Research on Design, Deployment, Automation, and Testing Strategies for 6G Mobile Core Network***
D. Satish Kumar (Nehru Institute of Engineering and Technology , India) G. Prabhakar (Thiagarajar College of Engineering, India) and R. Anand (Nehru Institute of Engineering and Technology, India)
Engineering Science Reference • © 2022 • 490pp • H/C (ISBN: 9781799896364) • US $360.00

***For an entire list of titles in this series, please visit:***
***http://www.igi-global.com/book-series/advances-wireless-technologies-telecommunication/73684***

701 East Chocolate Avenue, Hershey, PA 17033, USA
Tel: 717-533-8845 x100 • Fax: 717-533-8661
E-Mail: cust@igi-global.com • www.igi-global.com

# Table of Contents

# Detailed Table of Contents


*S. Vijay Anand, Sri Venkateswara College of Engineering, India*
*Sathis Kumar B., Vellore Institute of Technology, Chennai, India*

*Sudarson Rama Perumal, Rohini College of Engineering and Technology, India*
*Muthumanikandan V., Vellore Institute of Technology, Chennai, India*
*Sushmitha J., Rohini College of Engineering and Technology, India*

In recent decades, the most rapid change in wireless technology has been that flying ad hoc networks (FANETs) played a vital role in telecommunications. FANETs are flexible, inexpensive, and faster to deploy, which has led to the pathway to apply them in various applications such as military and civilian. However, FANETs have high mobility, and frequently changing topology patterns and tri-dimensional space movement make routing a challenging task in FANETs. FANETs differ from vehicular ad hoc network (VANETs) and mobile ad hoc networks (MANETs) in terms of features and attributes. It is always a challenge to choose the optimal path in any network using routing protocol. Due to these challenges, the performance and efficiency of the routing protocol have become critical. As network performance metrics like throughput, quality of service, user experience, response time, and other key parameters depends on the efficiency of the algorithm running inside the routing protocol, this chapter presents a novel routing protocol for FANETs in terms of distributed network routing algorithms and data forwarding routing.


*Sipra Swain, National Institute of Technology, Rourkela, India*
*Biswa Ranjan Senapati, ITER, Siksha 'O' Anusandhan University (Deemed), India*
*Pabitra Mohan Khilar, National Institute of Technology, Rourkela, India*

The demand for the quick transmission of data at any point and at any location motivates researchers from the industry and academics to work for the enhancement of ad hoc networks. With time, various forms of ad hoc networks are evolved. These are MANET, VANET, FANET, AANET, WSN, SPAN, etc. The initial objective of VANET is to provide safety applications by combining them with ITS. But later, the applications of VANET are extended to commercial, convenience, entertainment, and productive applications. Similarly, connections among multiple unmanned aerial vehicles (UAV) through wireless links, architectural simplicity, autonomous behaviour of UAV, etc. motivate the researchers to use FANET in various sectors like military, agriculture, and transportation for numerous applications. Search and rescue operations, forest fire detection and monitoring, crop management monitoring, area mapping, and road traffic monitoring are some of the applications of FANET. The authors mentioned some applications in the chapter using VANET, FANET, and the combination of VANET and FANET.



The tremendous evolution of wireless communication as well as the drastic adoption of technology by the latest computing devices known to be IoT, makes it possible for emerging applications to providing ubiquitous services. This technique transformed the quality of present lifestyle of the people. When compared with all other technologies, the mobile adhoc networks become widely adapted in many fields because of the non-requirement of centralized infrastructure support. Adopting this nature, it became easy to establish networks like WSN and also form networks using IoT devices. As FANET (flying/fast adhoc network) is known for its mobility and instant formation of network with the help of available nodes within its communication range, there is a great challenge related to mobility and authenticity of the participating devices by exempting malicious nodes. FANETs incorporate unmanned aerial vehicles and drones as a part of their communication networks. In this chapter, deployment of IoT-based FANETs along with mobility and security is handled.



There is great demand for VANETs in recent times. VANETs enable vehicular communication with the advent of latest trends in communication like 5G technology, software-defined networks, and fog and edge computing. Novel applications are evolving in recent times on VANETs with the proliferation of internet of things. Real test bed implementation is not always feasible with various limitations like expenditure and manpower and requires more time to experiment with the new facets of VANETs. Hence, the researchers should be aware of the variety of simulation tools that are capable of running VANET simulations. Simulation is a powerful tool in developing any critical/complex system that constitutes minimum cost and effort. The simulation tools of VANETs should support multiple mobility models, real-world communication protocols, and traffic modeling scenarios. This chapter gives a clear view on available tools and their characteristics on VANETs for research purposes.

# Preface

Editing a book is always challenging and great experience that also consumes a lot of time. We may not know many things in the initial stages and there is always a challenge to meet the outcomes at the end. Subsequently, you need to hurl yourself earnestly into something which isn't totally ensured.

When we write the proposal, we thought it would be a cake walk to edit the book with defined time schedules, but over the week's progress, we feel that FANETs and VANETs is a hot topic to deal with. We really had a challenge to find the experts working in these domains to get along.

So, when the opportunity was on us, we thought these were the sensations: doubts about the contents to include the authors to invite, their possible acceptance, the possible final result, the targets, and many other initial issues. Since the entire book was edited during the period of covid pandemic, we were discussing mostly online as we were located far away from each other. We both discuss many a times about the topics to selected for the book and finally boil down to find the applications, simulation scenarios of Vehicular Adhoc Networks (VANETs) and Flying Adhoc Networks (FANETs)

VANETs and FANETs are the future technologies that may rule the world as the increase in intelligent transportation systems (ITS) and drone-based applications like e commerce, agricultural monitoring, etc., were adopted by many corporate players. Hence this topic will really be an eye opener for the next generation to use the technologies like drones, UAVs, car infotainment systems, etc. This book will make the scholars, students and other industrial engineers to understand the variety of applications FANETs and VANETs were suitable and it will also tell the various simulation aspects of these networks and their characteristics.

This book is mainly intended for the following personnel in various industries, academia and research institutions:

- Students (Undergraduate, Graduate and Postgraduate)
- Research Scholars
- Faculty of various colleges and universities

- Engineers work in Automotive electronics
- Drone engineers
- Network engineers
- Consultants of various automotive technology companies.
- Academic body of many universities across the globe.

This book consists of 9 chapters that devotes mainly on the literature of Flying Adhoc Networks and Vehicular Adhoc networks, its applications and simulation aspect of these networks:

- **Chapter 1** introduces vehicular communication technology and focuses on developing, exploiting, managing, and securing them. The chapter also describes the most recent plans that have been implemented in a variety of application environments.
- **Chapter 2** discusses various issues and challenges while using flying Adhoc networks and it also presents a novel routing protocol for FANETs in terms of distributed network routing algorithms and data forwarding routing.
- **Chapter 3** describes the application scenarios of vehicular Adhoc networks, flyinf Adhoc networks and the combination of both these networks.
- **Chapter 4** present and discuss the challenges related to mobility and security threats in FANETs along with the applications of FANETs using Internet of Things.
- **Chapter 5** present the characteristics of VANETs and the available simulation tools. Most of these tools were either event driven or time driven and these tools models the given vehicular networks. This chapter also mentions the tools that are good for traffic modeling, plotting the characteristics, traffic generation, etc.
- **Chapter 6** present two case studies: User defined road structure and Roads designed through open Street Maps to showcase a real time simulation of VANET using Simulation of Urban Mobility (SUMO) and the chapter end with a discussion of simulation and analyses of various network performance metrics like throughput, PHY overheads and packet delivery ratio, etc.
- **Chapter 7** presents and discusses about a new subset of flying Adhoc networks called as Internet of Flying Things (IoFT) which are suitable for applications that involves sensors and actuators for capturing the environment.
- **Chapter 8** presents threats and attacks in VANETS and motivates the need for developing the robust IOV system which ensure the secure infrastructure.
- **Chapter 9** present enhanced blockchain data model, in which new timestamps are added to the blockchain named BiTchain, and the chapter also introduce an expressive temporal query language, named BiTEQL. This chapter also

handles how these blockchain based applications are suitable for Flying Adhoc networks and vehicular Adhoc networks.

To be said precisely, this book answers the following questions raised by many scholars and students

- *What are the differences between flying Adhoc networks and vehicular Adhoc networks?*
- *How do I generate a vehicular traffic?*
- *How the performance metrics can be measured for a FANET and VANET?*
- *What is the state of art in the deployment of UAVs and Drones?*
- *How do I customize drones for different applications?*
- *In what way, the vehicular networks and the flying networks are secure and how they can prevent attacks during communication?*

There are many more questions will be answered in the book. No doubt, in the years to come, Flying Networks and Vehicular networks will be deployed as much as in many applications related to military, surveillance, e commerce, agriculture, etc. In the researchers' perspective also, there are plenty of open problems to solve for the society and the world. This book will surely help those researchers to start with what is FANETs and VANETs and how research in these domains can be helpful.

We really feel and hope that this book may help the students, scholars and acquaint new followers with the technology and the topic.

We wish you a pleasant reading.

*T. S. Pradeep Kumar*
*Vellore Institute of Technology, India*

*M. Alamelu*
*Kumaraguru College of Technology, India*

# Acknowledgment

Taking up a book project is harder than we thought since this is our first book ever to be edited and more rewarding than we could have ever imagined. None of this would have been possible without the support of our friends, family members, and academic connections.

We would like to acknowledge the help of all the people involved in this project and, more specifically, to the authors and reviewers who took part in the review process. Without their support, this book would not have become a reality.

First, we would like to thank each one of the authors for their contributions. Our sincere gratitude goes to the chapter's authors who contributed their time and expertise to this book. We thank all the authors of the chapters for their commitment to this endeavor and their timely response to our incessant requests for revisions.

Second, the editors wish to acknowledge the valuable contributions of the reviewers regarding the improvement of quality, coherence, and content presentation of chapters. Most of the authors also served as referees; we highly appreciate their double task.

The editors would like to recognize the contributions of editorial board in shaping the nature of the chapters in this book. In addition, we wish to thank the editorial staff at IGI Global for their professional assistance and patience.

Without the support from our Management of Vellore Institute of Technology (VIT Chennai), Kumaraguru College of Technology (KCT), academic heads, colleagues, peers of VIT Chennai and Kumaraguru College of Technology, Coimbatore, this book would not exist. A sincere thanks to each one of them.

We want to thank God most of all, because without God we wouldn't be able to do this.

# Introduction

The proliferation of communication, Internet technologies from the past few years has been phenomenal. They have now emerged as the choice of fast-moving ad hoc communications due to their feasible, convenient and low expenditure features. FANETs are a group of small Unmanned Aerial Vehicles (UAVs) that are integrated to achieve high level goals. VANETs are a group of mobile or stationary vehicles connected wirelessly. The main use of VANETs was to provide safety and comfort to drivers in vehicular environments.

This book incorporates all the features, tools, history, challenges, sample simulation modelling and open research issues in FANETs and VANETs.

The book does not require prior knowledge of VANETs and FANETs. This book can be understood without any prerequisites of ad hoc networks.

The following is a brief description of the topics covered in each chapter.

Chapter 1 deals with various challenges in VANET. VANETs is a rapidly developing field of vehicular communication technology. The main purpose of this technology is to create a safe and secure environment for vehicles. This chapter highlights recent issues such as developments, exploitation, safety, and security issues in VANETs.

Chapter 2 introduces the routing constraints and challenges in FANETs. Efficient routing is very crucial in FANETs to maintain QoS. Various applications, evolution of drones, UAVs and how routing should be done in FANETs is presented in this chapter.

Chapter 3 introduces the evolution of VANET and FANET for real life applications. This chapter also projects how VANET and FANET can be integrated for various applications.

Chapter 4 presents Establishment of FANETs using IoT based UAV. All the issues regarding mobility and authenticity were discussed in this chapter.

Chapter 5 gives a clear view on available tools and their characteristics on VANETs for research purposes. How to create traffic scenarios in SUMO for VANETs is illustrated in this chapter.

Chapter 6 presents real time simulation of vehicular Ad hoc networks using Simulation of Urban Mobility (SUMO) in User defined road structure and Roads designed through open Street Maps. This chapter provides a brief introduction on how to analyze the various network performance metrics like throughput, phy overheads and packet delivery ratio using the simulation.

Chapter 7 discusses various FANET applications. FANETs features like ease of deployment, dynamic configuration, low maintenance costs, high mobility, and faster reaction were explained in this chapter.

Chapter 8 explains how IoT and security mechanisms are incorporated with the VANET system to design the robust framework. This chapter gives the analysis and recommendation for developing the robust IOV system which ensures the secure infrastructure.

Chapter 9 introduces an expressive temporal query language, named BiTEQL and enhanced block chain model BiTchain. This chapter gives an overview on how to add new timestamps to the blockchain, which are assigned by data originators to represent the valid time of data recorded within a transaction

I take this opportunity to thank my family, many of my colleagues and readers for their suggestions and comments, that helped me to improve the quality of the book.

# Chapter 1
# Challenges in VANET

**S. Vijay Anand**
*Sri Venkateswara College of Engineering, India*

**Sathis Kumar B.**
*Vellore Institute of Technology, Chennai, India*

## ABSTRACT

*The vehicular ad-hoc network (VANET) is a wide and rapidly developing field of vehicular communication technology study. A VANET is a network with no infrastructure. It is used to improve safety in terms of applications and convenience of use while driving. VANET is a vehicle-to-vehicle network that allows automobiles to share secure data while travelling on highways or roads. VANET applications are being developed for cities throughout the world. VANET delivers an identity recognition technology that has a significant impact on improving activity administrations and reducing traffic accidents. The main purpose of this technology is to create a safe and secure environment for vehicles. Many architectures, algorithms, and protocols have been developed and implemented in recent years to improve the performance of automobiles while travelling. The writers of this research study highlight recent issues such as developments, exploitation, safety, and security issues, as well as the most recent plans that have been run in various situations.*

DOI: 10.4018/978-1-6684-3610-3.ch001

## INTRODUCTION

Vehicular Adhoc Networks (VANETs) are special type of MANETs (Mobile Adhoc Networks) in which the vehicles are nodes that act as a participating node and as well as a router. Since vehicles can send messages to other vehicles in the networks, there need not be any network infrastructure. But however, there may be roadside assistant networks that help vehicles in diagnostics, accident rescue, etc.

The main characteristics of VANETs are higher node mobility and Speed of the nodes (vehicles). The protocols that were used in VANETs in the earlier days are DSRC (Dedicated Short Range communication) which has so many problems like interference, etc. Now the current research in VANETs leads to the use of MANETs protocols for VANETs (Meneguette et al., 2015).

The Dynamic Source routing (DSR) and Adhoc on Demand Distance Vector (AODV) suits well for the VANETs as these protocols are suitable for multi hop wireless ad hoc networks, which is the prime requirement of VANETs. There are many literatures that covers the importance of VANETs, their deployments, their applications and metrics of other fast moving Adhoc networks as well. Some of these literatures are suggested and experimented as follows: (Anjum et al.,2020) designed a radio frequency-based sensor network for energy harvesting and hence the optimal energy is being used while transmitting and receiving. This paper uses a model that uses reward allocation when the states transition happens. This work is extended to handle optimal energy management for Internet of Vehicles as well.

(Kumar & Krishna, 2018) suggested a model that is solved using reinforcement learning which can optimize the power usage in internet of things and internet of vehicles. This paper identifies the power profile of various devices in the system and based on the power profile, the system is modeled using the semi-Markov decision process (SMDP) and solved using the reinforcement learning. DSRC protocol is standardized to work under the frequency range of 5.9GHz along with the WAVE 1609 standard as mentioned by (Maddio et al.,2013) in their work. Due to the size of vehicle density and speed of the vehicles, the DSRC performance analysis is highly complicated.

(Su & Zhang 2007) proposed a medium access control protocol that can send safety messages from cluster to cluster. They use content free and contention-based protocol within clusters and between cluster heads respectively. Each cluster heads can relay the safety message in real time to other cluster heads and the heads in turn send those messages to the cluster vehicles. The relay of these messages can be sent to both real time traffic and non-real time traffic as well.

(Zhang et al.,2019a) proposed a medium access Control protocol based on the DSRC protocol. This protocol is well suited for the purpose of basic safety messages which is been sent to nearby vehicles during a collision or any accidents. This protocol is designed in a such a way that it obeys the IEEE 802.11p protocol as well. (Zhang et al.,2019b) proposed an analytical evaluation of increasing the packet delivery ratio of the vehicular network through a new hybrid MAC protocol. The hybrid mac protocol handles the packets and the transmission in one dissemination period so that the delivery latency is too small. (Wu et al., 2009) implemented and suggested a new energy efficient protocol that can minimize the energy consumption of the DSRC protocol to 44%. This protocol divides the vehicles into many clusters and within each cluster, there will be a different wake-up/sleep schedule so that the average energy consumption will be minimized. This protocol is based on the IEEE 802.11 power saving mechanism which is then integrated in to DSRC protocol for energy optimization.

In this chapter, the topics were divided into many sections related to VANETs properties like Volatility, time delay, networking standards, node mobility and security requirements.

## Volatility

VANET (vehicular ad hoc network) is a highly volatile network due to the high speed of vehicles. The connection between vehicles lasts only a few seconds. The activity of connecting and disconnecting nodes is continuous. Furthermore, nodes are moving in opposite directions, increasing the network's volatility. VANET faces the following challenges as a result of its volatility (Rehman et al.,2013).

### Maintain Communication on Move

Because of the network's volatility, nodes in the network move around at random and lose connections that have been established for communication. Disconnection of nodes interrupts the transmission of previous data packets. If a node's connection is lost, it's critical to create a new path between the source and the destination. In dynamic networks like VANET, it adds extra overhead.

As a result, maintaining communication between moving nodes is a difficult task.

### The Message Should Be Delivered on Time

For a brief period of time, VANET nodes remain in close proximity to one another. As a result, the time between the source sending the message and the receiver receiving it should be as short as possible. Achieving low latency in vehicular communication is always a difficult task.

### Routing at High Speed

Routing is the process of determining a communication path between a source and a destination. It's difficult to adapt existing wireless routing protocols to the VANET environment because nodes are moving. Rerouting is frequently triggered by the connection and disconnection of nodes. Fast routing is critical for successful communication and avoiding squandering the effort already expended on communication.

### Alternate Communication Path Selection in Real Time

Due to the volatility of the VANET, the rerouting process is frequently initiated. Finding an alternative communication path among the available options is difficult because it takes into account not only the distance between nodes but also the velocity, direction, and density of nodes. As a result, real-time path selection in the VANET is a research topic.

## SIGNIFICANT TIME DELAY IN MESSAGES DELIVERY

The time interval between the source node sending messages and the receiver node receiving messages is known as latency. Highly mobile nodes, such as vehicles, in a VANET may be travelling in opposite directions. The nodes are only in close proximity to each other for very short periods of time. It is critical that the message be received by the destination vehicle within a certain amount of time (Liu et al., 2011).

The network must meet the critical latency requirement in order to achieve this communication. The VANET network's communication should have a low latency.

In a VANET, achieving low latency while communicating is a difficult task. The vehicular network's primary goal is to transmit safety messages. Safety messages should always take precedence, and they should arrive at their destination on time. To implement this feature, a network delay must be overcome.

The following are the main obstacles to overcoming a delay in vehicular networks:

- Changing neighbourhoods frequently due to high mobility
- Significantly increasing network load due to high-density environment
- Packet dropping due to exposed and hidden terminal problems
- Connectivity issue caused by variations in received signal power

## ESSENTIAL CHANGE IN VEHICLES AND ROADS

Any country's road network is critical to the development of that country. Every year, the number of roads in developing countries grows dramatically.

Millions of new vehicles are added to the network every year. The VANET faces the following challenges as the number of vehicles and roads grows.

### To Construct the Necessary Infrastructure for New Roads

All necessary devices, such as roadside units (RSUs) and servers, must be deployed to provide VANET services on newly established roads. It is difficult to do so in order to drastically increase infrastructure. It requires both resources and deployment costs.

### New Vehicle Registration and Credentials Generation

Every new vehicle is registered with a reputable organisation (TA). The TA issues certificates or other credentials to vehicles in order for them to access the network. The TA assigns each vehicle a unique identification number. The task of registering and assigning credentials to the growing number of vehicles is difficult. The TA keeps these credentials up to date throughout the life of the vehicle. Monitoring vehicles and providing some common services over the network are becoming more difficult tasks as the number of vehicles grows (Guo et al.,2017).

### Decrease the Network's Performance as the Number of Vehicles on the Road Grows

The increasing number of vehicles places additional strain on the existing network. When an infrastructure is overburdened due to a high density of vehicles, the network's performance suffers. It's possible that the network will slow down. To achieve the required performance, network resource capacity must be updated.

## NETWORKING STANDARDS FROM AROUND THE WORLD

An ad hoc network is referred to as a VANET. The communication requirements for VANETs are different from those for other ad hoc networks. The protocols and standards that are available for vehicular networks do not apply to VANET. It is necessary to adapt these standards to meet new VANET requirements, or to create new standards. Adapting existing standards and creating new ones are both difficult tasks. Choosing a protocol from a large list of standard protocols is a difficult task.

## NODE MOBILITY IS EXTREMELY HIGH

The nodes in a vehicular network, as we all know, are high-speed vehicles. It's a network with a lot of mobility. The following network challenges are caused by node mobility (Muniyandi et al.,2020).

### Network Node Disconnection on a Regular Basis

Because nodes move at different speeds, they frequently connect and disconnect. Communication is disrupted when nodes are disconnected. As a result, data must be retransmitted. At the receiver end, disconnection can result in data loss and a time delay in sending data. The network's performance suffers as a result. Providing network rerouting and overcoming the disconnection effect is a difficult task.

### Topology Changes Frequently

The mobility of nodes causes the topology to change frequently.

### Retransmission of Data of a Communication

Failure occurs, data must be retransmitted to the receiver node. Retransmission of data necessitates additional overhead. It consumes network resources as well as bandwidth. It also causes a delay in message delivery.

### Packet Delivery Ratio Is Low

The packet delivery ratio (PDR) is a metric used to assess network performance. It's the difference between the number of packets received at the receiver end and the number of packets sent by the source node. The PDR is negatively affected by the high mobility of nodes, as communication is frequently interrupted.

## NETWORK SECURITY

A vehicle in a VANET is nothing more than an intelligent mobile node that communicates with its network neighbours, which can be other vehicles or RSUs.

We discussed various approaches for improving the overall security, robustness, and efficiency of the VANET system ((Pavithra & Nagabhushana,2020).

## Security Requirements for VANET

1. **Accreditation:** Certification/authentication verifies the source node and ensures that only valid nodes can send messages, reducing the attackers' intent significantly. The authentication method should be able to: The authentication method should be able to:
   a. Authenticate nodes
   b. Protect nodes' privacy
   c. Be very efficient and quick.
2. **Message integrity:** Message integrity refers to the fact that messages are not changed in the middle of network transits. When it comes to message integrity, it should be ensured that any message received by any driver is not false or incorrect.
3. **Data non-repudiation:** Message/data non-repudiation ensures that the sender cannot deny a message sent by the same node.

While doing so, any non-repudiation of messages should be checked by only authorised authorities, not by anyone.

4. **Object/entity authentication:** Entity authentication ensures that the message sender is the same as the message sender specified in the message. And that the sender can be identified in a short period of time in a specific location.
5. **Data/message confidentiality:** Message confidentiality ensures that only authenticated and private nodes can receive and send messages. That message is hidden from third parties.
6. **Security:** Security ensures that messages are only accessed and viewed by authenticated users, and that third parties or unauthorised users are unable to view them. Other nodes can also track the location of vehicles and submit queries to authorities.
7. **Real-time assurances:** In any VANET real-time application, time assurance is critical. To ensure time sensitivity, protocols, methods, or applications can be used.
8. **Time-optimal security check:** A VANET is a dense network with a large number of nodes clustered together in a small area. And all of those nodes should be authenticated in a short period of time. It is critical to ensure quick authentication in order for authorities to serve more vehicles.

## Issues With VANET Security

Different security breaches, which violate the aforementioned security requirements, can be defined as VANET security and VANET problems. These security breaches occurred for a variety of reasons, which can be divided into three categories:

## Threats to Security

### Security Concerns

- Authentication and privacy: A message should be verified, and its node authenticated before it is delivered to its intended destination. While doing so, the privacy of both nodes and messages should be maintained.
- High vehicle density and mobility: RSU and authority servers are under a lot of strain due to the large number of vehicles, messages, and their authentication. As a result, authentication time should be kept as short as possible so that more vehicles and messages can be served in the same amount of time.
- Message delivery in real time: Because VANET applications are primarily used to notify events such as accidents, traffic, and weather conditions, they must adhere to a strict message delivery time and prioritise the messages.
- Location awareness: Knowing the location of the original node can help identify attacks like Sybil, and many applications require vehicle location, which should be tracked to provide better services.
- Attacks
- Greedy drivers or businesses: Most drivers are honest and follow traffic routes, but there are exceptions where drivers want to maximise available resources to gain a competitive advantage in business or for personal reasons. For example, drivers may send out false information about accidents or hazardous conditions, causing other vehicles to divert their traffic, allowing those greedy drivers to use those routes to their full potential.

Message delay is a type of attack that occurs when required messages must be delivered on time in order to avoid network issues such as accidents or traffic congestion.

- Eavesdroppers/snoopers: These individuals attempt to obtain unauthorised additional information about others. They can obtain the identity of another vehicle or hack the system to obtain the information they require.
- Pranksters: Pranksters are mostly teenagers who, for fun, send incorrect messages to other vehicles, causing them to become misguided and end up

in dangerous situations. They also use Distributed Denial of Service (DDoS) attacks to delay the delivery of messages.

- Insiders in the industry: Hardware manufacturers install firmware that compromises security and steals information. This type of security breach is difficult to detect and dangerous to the VANET. Using tampering devices, an attacker can also steal the identity and keys of another vehicle. An intruder in their network can compromise the security and gain access to confidential information.
- Assailants who are malicious: The most serious concern for our security system is that attackers harm the vehicular network.

These attackers are more experienced and are aware of the system's flaws. They can use techniques such as the Sybil attack, replay attack, DDoS attack, and false messages to attack.

## Properties

In this network, VANET has very unique security characteristics.

- Vehicles have their own batteries and OBUs (on- board units) have a lot of processing power.

In comparison to MANET (mobile ad hoc network), where the most challenging issue is power shortage, taking security measures and implementing security checks/ algorithms is somewhat appropriate.

- Capable of tracking location and time:

RSU and TA are in constant communication, and it is simple to track its current location using GPS and time.

Several algorithms can be implemented using this location and time information to track malicious nodes, the Sybil attack, and so on.

- Periodic maintenance and inspection: Authorized agencies can check for firmware software on a regular basis to avoid privacy concerns. Authorities also change OBU's key and other secure information on a regular basis.

In comparison to VANET, there is no provision for a centralised certificate authority in MANET.

Each vehicle should be registered with centralised agencies and given a unique TA identification number.

- Existing law and infrastructure: Existing law allows attackers to avoid security breaches, and current law can catch and punish them if any attacks are discovered in the system.

## VANET Security Critical Components

- Vehicle black box: A tamper-resistant device that stores vehicles as well as all critical electronic data such as speed, time, messages, and location.
- Trusted component (TC): Cryptographic keys are critical to the security of cryptographic operations, and storing keys requires proper hardware protection. The term "trusted component" refers to secure hardware that aids in the storage of cryptographic keys in a secure vault.
- Vehicle identification: Vehicle identification is provided in the form of electronic licence plates, which helps to automate vehicle inspection and verification.

Management system for key: In the key management system (KMS), the TA is responsible for issuing credential pairs such as public and private keys to all registered vehicles. There will be several TAs corresponding to different regions in the case of a larger region; TAs will verify each other's authenticity (Krishnan & Kumar 2020).

## CONCLUSION

VANET being a wellbeing data sharing medium, needs secure and safe climate. VANET has exceptionally wide scope for assaults because of its profoundly powerful nature, remote mode of correspondence and every now and again evolving topology. Security issues and difficulties connected with VANET profoundly affect productive usefulness of the system. Today, VANET are broadly conveyed because of its improving highlights of giving protected, secure and comfort driving VANET, elements of VANET, need of safety in VANET are the hotly debated issues connected with the ongoing situation. In this paper, we have done a writing study about different kinds of assaults, their preventive measures, sort of aggressors furthermore, some current security answer for assaults in VANET.

## REFERENCES

Anjum, S. S., Noor, R., Ahmedy, I., Anisi, M. H., Azzuhri, S. R., Kiah, M. L. M., ... Kumar, P. (2020). An Optimal Management Modelling of Energy Harvesting and Transfer for IoT-based RF-enabled Sensor Networks. *Ad-Hoc & Sensor Wireless Networks*, 46.

Guo, L., Dong, M., Ota, K., Li, Q., Ye, T., Wu, J., & Li, J. (2017, April). A secure mechanism for big data collect in in large scale internet of vehicle. *IEEE Internet of Things Journal*, *4*(2), 601610.

Krishnan & Kumar. (2020). Security and Privacy in VANET: Concepts, Solutions and Challenges. *2020 International Conference on Inventive Computation Technologies (ICICT)*, 789-794. doi: 10.1109/ICICT48043.2020.9112535

Kumar, T. P., & Krishna, P. V. (2018). Power modelling of sensors for IoT using reinforcement learning. *International Journal of Advanced Intelligence Paradigms*, *10*(1-2), 3–22.

Liu, Y., Niu, J., Qu, G., Cai, Q., & Ma, J. (2011). Message delivery delay analysis in VANETs with a bidirectional traffic model. *2011 7th International Wireless Communications and Mobile Computing Conference*, 1754-1759. doi: 10.1109/IWCMC.2011.5982801

Maddio, S., Cidronali, A., Palonghi, A., & Manes, G. (2013, June). A reconfigurable leakage canceler at 5.8 GHz for DSRC applications. In *2013 IEEE MTT-S International Microwave Symposium Digest (MTT)* (pp. 1-3). IEEE.

Meneguette, R. I., Bittencourt, L. F., & Madeira, E. R. M. (2013). A seamless flow mobility management architecture for vehicular communication networks. *Journal of Communications and Networks (Seoul)*, *15*(2), 207–216. doi:10.1109/JCN.2013.000034

Muniyandi, R.,, Qamar, F.,, & Jasim, Naeem, A. (2020). Genetic Optimized Location Aided Routing Protocol for VANET Based on Rectangular Estimation of Position. *Applied Sciences (Basel, Switzerland)*, *10*, 5759. doi:10.3390/app10175759

Pavithra, T., & Nagabhushana, B. S. (2020). A Survey on Security in VANETs. *2020 Second International Conference on Inventive Research in Computing Applications (ICIRCA)*, 881-889. doi: 10.1109/ICIRCA48905.2020.9182823

Su, H., & Zhang, X. (2007). Clustering-based multichannel MAC protocols for QoS provisionings over vehicular ad hoc networks. *IEEE Transactions on Vehicular Technology*, *56*(6), 3309–3323.

ur Rehman, Khan, Zia, & Zheng. (2013). Vehicular Ad -Hoc Networks (VANETs) – An Overview and Challenges. *JWNC*.

Wu, S. H., Chen, C. M., & Chen, M. S. (2009). An asymmetric and asynchronous energy conservation protocol for vehicular networks. *IEEE Transactions on Mobile Computing*, *9*(1), 98–111.

Zhang, M., Ali, G. M. N., Chong, P. H. J., Seet, B. C., & Kumar, A. (2019). A novel hybrid mac protocol for basic safety message broadcasting in vehicular networks. *IEEE Transactions on Intelligent Transportation Systems*, *21*(10), 4269–4282.

Zhang, M., Chong, P. H. J., & Seet, B. C. (2019). Performance analysis and boost for a MAC protocol in vehicular networks. *IEEE Transactions on Vehicular Technology*, *68*(9), 8721–8728.

# Chapter 2
# Flying Ad hoc Networks Routing Constraints and Challenge Perspectives

**Sudarson Rama Perumal**
*Rohini College of Engineering and Technology, India*

**Muthumanikandan V.**
https://orcid.org/0000-0002-5863-5047
*Vellore Institute of Technology, Chennai, India*

**Sushmitha J.**
*Rohini College of Engineering and Technology, India*

## ABSTRACT

*In recent decades, the most rapid change in wireless technology has been that flying ad hoc networks (FANETs) played a vital role in telecommunications. FANETs are flexible, inexpensive, and faster to deploy, which has led to the pathway to apply them in various applications such as military and civilian. However, FANETs have high mobility, and frequently changing topology patterns and tri-dimensional space movement make routing a challenging task in FANETs. FANETs differ from vehicular ad hoc network (VANETs) and mobile ad hoc networks (MANETs) in terms of features and attributes. It is always a challenge to choose the optimal path in any network using routing protocol. Due to these challenges, the performance and efficiency of the routing protocol have become critical. As network performance metrics like throughput, quality of service, user experience, response time, and other key parameters depends on the efficiency of the algorithm running inside the routing protocol, this chapter presents a novel routing protocol for FANETs in terms of distributed network routing algorithms and data forwarding routing.*

DOI: 10.4018/978-1-6684-3610-3.ch002

## INTRODUCTION

FANETs are now a reality thanks to advancements in wireless communication technology that have played a key role. Global 5G wireless network infrastructure revenue is expected to reach $4.2 billion by 2021, an increase of 89% over the revenue generated in 2020, (Zhao, H. et.al., 2018). It is clear from these numbers that wireless technologies will be ideal for FANET applications since they will give more coverage and faster speeds. Flying Adhoc Nodes are quick, maneuverable, and complicated in their flight environment and high degree of combat. P2P and MPR are the most common methods for establishing communication between nodes as proposed (Gong, J. et.al., 2018). When two nodes are in close proximity to each other, point-to-point communication is possible. However, if the nodes are too far apart, MPR techniques are used in their place. (Park, S. Y et.al 2018) proposed a packet forwarder or a relay agent can be established at any intermediate node in MPR.

There has been a lot of research in these areas in the last two decades, especially in the MANET and VANET areas (Lu, J., et.al., 2018). Small Unmanned Aerial Vehicles (UAVs) in a Flying Adhoc Network (FANET) have recently attracted attention because of their availability, versatility, adaptability, autonomy, and ease of deployment (Thammawichai, M., et.al., 2017).

It is also possible to use UAVs in a wide range of applications because of their huge coverage and ease of installation. However, there are several concerns that need to be addressed, such as the high mobility and sparse deployment of UAVs (Trotta, A. et.al., 2018).

## MOTIVATION

To keep pace with FANET's ever-changing requirements, a routing protocol must be flexible enough to accommodate its highly mobile and dynamic nature. Packet loss, delay, and jitter have a significant impact on a network's ability to provide a high level of service. Second, it must be scalable in order to deliver an appropriate degree of throughput against the network demand. Nodes must be able to preserve energy in order to extend network life spans, which is a third and most essential issue (Wang, Y., et.al., 2018).

*Figure 1. (a) One to One UAV network; (b) Many-UAV network.*

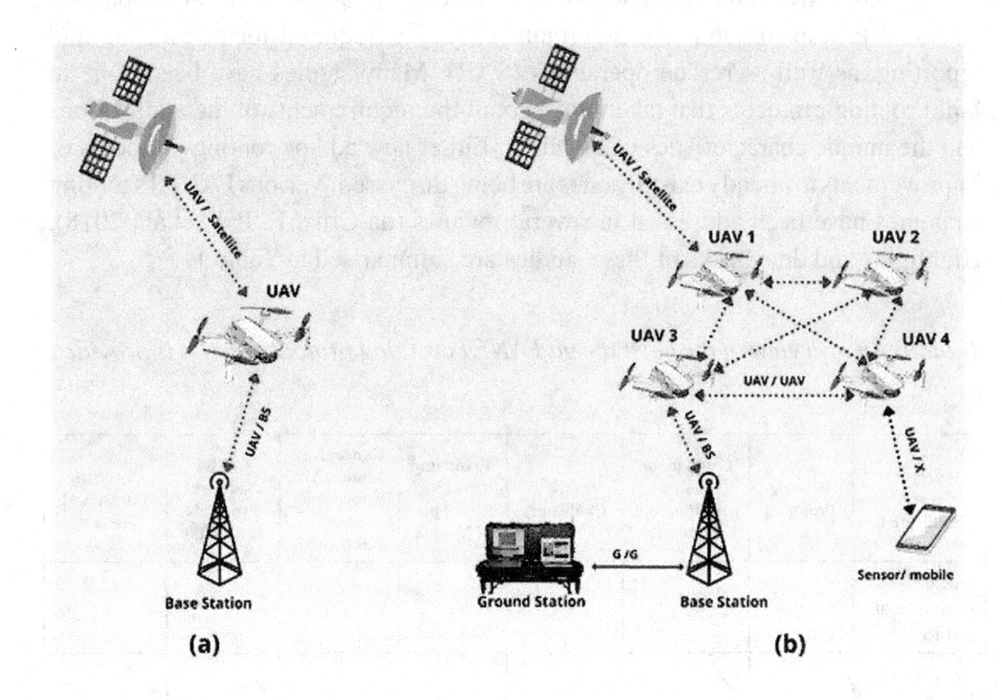


## UAV CATEGORIZATION

Rising unmanned aerial vehicles (UAVs), moderate unmanned aerial vehicles (MAUs), and low-level unmanned aerial vehicles (LAUs) are naturally categorized in FANETs depending on their height. HAUs, comprising satellite, aircrafts, and hot air balloons, are at heights more than 20 kilometers and are virtually completely stationary in their flight path. (Zheng, Z. et.al., 2018), MAUs soar at moderate altitudes up to 11 kilometers above the ground, similar to aircraft, and move more swiftly in the sight of ground nodes. However, this has changed recently. Certain unmanned aerial vehicles (UAVs) have been categorized based on a variety of qualities in order to communicate with base stations, land node (e.g., cars or ships), and spacecraft. Data transmission between UAV nodes requires the use of a routing mechanism. In many cases, VANET and MANET-specific ad hoc network routing protocols fail to meet the needs of FANETs (Singh, K., et.al., 2019).

Developing successful routing protocols for FANETs is difficult because of the unique properties of FANETs, such as flying in three dimensions, a low node density, fast topology changes, broken links, network segmentation, and a lack of resources (Wang, H., et.al., 2018).

In the context of FANET applications, service quality (QoS) is a crucial component. Actual data transmission with minimum latency is required for monitoring and reporting, as well as rescue operations (SAR). Many studies have been done to build routing protocols that take into account the requirements of the applications and the unique characteristics of FANETs. Either new ad hoc routing protocols or improvements to already existing ones are being discussed. Various FANETs routing strategies have been addressed in several reviews (da Cruz, E. P. F. et.al., 2018). Highlights and drawbacks of these studies are summarised in Table 1.

*Table 1. An overview of the most recent FANETs routing protocol review is provided below.*

| Reference/ Year of Publication | Routing Protocols | Comparison Analysis of Routing Protocols | Routing Challenges | Taxonomy of Concepts of Motion | Concepts of Motion Are Compared and Contrasted. | FANET Network Protocols (FANETs) | Currently Unresolved Issues |
|---|---|---|---|---|---|---|---|
| Sánchez-García, et al, 2014 | ✓ | x | x | x | x | ✓ | ✓ |
| Oubbati, O. S et al., | ✓ | ✓ | x | x | x | ✓ | ✓ |
| Sharma, V et al., | ✓ | ✓ | ✓ | x | x | ✓ | ✓ |
| Altawy, R et al., | ✓ | x | x | x | x | ✓ | ✓ |
| Fan, X et al., | ✓ | ✓ | ✓ | x | x | x | ✓ |
| Khan, M. A et al., | ✓ | ✓ | x | x | x | ✓ | ✓ |
| Maxa, J. A et al., | ✓ | x | ✓ | x | x | ✓ | ✓ |
| J Wang et al., | ✓ | ✓ | x | x | x | ✓ | ✓ |
| Hayat, S et al., | ✓ | ✓ | x | x | x | x | ✓ |
| This review | ✓ | ✓ | ✓ | ✓ | ✓ | ✓ | ✓ |

## FANET ARCHITECTURE

FANET is a network of unmanned aerial vehicles (UAVs) and ground-based stations (GBSs) that are linked together on their own. UAVs and ground-based systems (GBS) must both participate in the transmission process, which must be performed on an individual basis (Khan, M. A., et.al., 2018). UAVs having extra characteristics are often selected by this network to act as bridges connecting GBSs as well as other UAVs, therefore increasing wireless coverage in overall. There are also a variety of communication methods, network structures, and Numerous types need different qualities. Campion, M., et.al., 2018). This chapter has been divided into three sub groups in order to achieve this purpose. Lastly, the specific properties and features of FANETs are discussed in depth here. Figure 2 serves as a constant reference point throughout the next sections (Jiang, J., et.al., 2018).

Creating a fully cooperative FANET system necessitates the development of a set of processes and rules for exchanging information between GBSs and UAVs. The communication groups that are employed will depend on the applications that UAVs will be using. Moreover, in our evaluation, we bring out their respective advantages and disadvantages (Sánchez-García, J., et.al., 2018).

## Management With Just a Centrally Controlled Structure

All UAVs are strongly linked to one or even more GBSs, which are capable of communicating with all of the UAVs at the same time. GBSs are required to route all data traffic because Inter-UAV connectivity is not feasible at this time. (Cumino, P., et.al., 2018.) Many advantages can be gained from such an organization, including greater reusability in the case of a breakdown of an unmanned aerial vehicle simultaneous task and enhanced calculation and storage capacities. Last but not least, the GBS is a single point of failure that can cause the entire network to go down in the event of a failure or assault (Tareque, M. H., et.al., 2015).

## Multi-Group Organization

As long as the centralized structure is maintained, the UAVs can communicate with each other in an unorthodox manner. UAVs are allocated to serve as gateways between groups and the GBSs in numerous groups, each with a dedicated UAV.The GBSs are used for inter-group communications, however the GBSs are not used for internal communications (Arafat, M. Y., et.al., 2018). This organization is more efficient than a centralized one that supports a massive number of unmanned aerial vehicles (UAVs) with various transmission and flying skills Yet, As an added risk, the failure of a specific GBS may lead to network partition, isolating a particular UAV fleet from other UAVs on the network (Oubbati, O. S., et.al., 2017).

## Cellular Organization

GBS units that transport unmanned aerial vehicles (UAVs) offer a viable alternative for enabling the installation of a wide range of commercial and defense purposes simpler. Cells can give a significant amount of transmission area in a certain location when they work together. Such an organization permits UAVs to interact directly with others or via GBSs (Global Broadcast System). It is also vulnerable because of the static GBSs that may break at any moment, leading in total losing control of many or all Aircraft Many issues must be addressed before cell phone networks are widely implemented (Sharma, V., et.al., 2017).

## FANET COMMUNICATION

Each FANET node (i.e., UAV, GBS, and satellites) can serve as a final solution, according to the aforementioned FANET organizations. To handle the frequent topological changes, all these nodes can work together as relays and form a network. Aside from the two types of ground-to-ground communication, there are three types of communication that need to be considered in FANETs: In the next sections, we'll go into greater depth about these communications (Altawy, R., et.al., 2016).

### UAV-to-UAV (U2U) Communication

UAVs connect directly with one other to meet the demands of various missions by exchanging data packets on a regular basis. Due to bandwidth range limits, the inter process carried out via the use of other unmanned aerial vehicles. This is critical if you want to expand the coverage of a given topic. Because there are no barriers between UAVs in the sky, line-of-sight (LoS) is the most common mode of communication in U2U. But, there were also times when boundary is not guaranteed, including when UAVs hover near tall objects or hills (Fan, X., et.al., 2017).

### UAV-to-Ground (U2G) Communication

GBS systems are set up on the floor to communicate crucial control information. and command messages for improved control of flying UAVs. Additionally, GBSs are utilised to connect various groups of UAVs together. In general, certain UAVs may communicate with GBSs to minimise network cost and increase efficiency and connectivity. Because of the presence of impediments on the ground that generate echoes and scattering effects, UAVs are unable to provide a Line of Sight with GBSs at lower heights. (Khan, M. A., et.al., 2017).

## Satellite Technology

UAVs are frequently used in difficult-to-install GBS locations, including as the ocean and hilly terrain. Satellites may be a viable solution for centrally commanding UAVs while simultaneously providing critical LoS coverage, resulting in the establishment of Satellite Communication (SATCOM). Both the interchange of important data among UAVs and the delivery of gathered data to a broadcaster positioned far distant on the ground benefit from SATCOM. Nevertheless, it was not an expense alternative (Maxa, J. A., et.al., 2017).

*Table 2. Comparison of Fanet Communications*

| | U2U | U2G | SATCOM |
|---|---|---|---|
| LoS | High | Medium | High |
| Cost | Cheaper | Expensive | Highly expensive |
| Coverage | Medium | Large | High |
| Exploitation | Short-term | Mid-term | Long-term |

## Attributes of Routing

In depending on the number of UAVs, flying duration, and communication limits, each project or program has its own set of requirements. This distinguishes FANETs from other types of ad hoc networks by diversifying their properties and making them distinctive. We present a full overview of the most important FANET features that are taken into account during the implementation of these kind of networks in this chapter.

## Rate of Access Point

LAUs are one of the most extensively explored varieties of UAVs. In this categorization, we differentiate between two types of unmanned aerial vehicles (UAVs): I UAVs with rotational rotors (RW) and (ii) Unmanned aerial with fixed - wing aircraft (FW). In principle, both types of Uavs are planted in two and three (2D or 3D) space, and their mobility is controlled by the mission. Their motions are really quite dynamic, tend to range from remain totally still (as in aerial observation or coverage) to trying to fly at high throttle (e.g., in deliverance of products or disaster response task) (Wang, J., et.al., 2017).

*Table 3. Uavs with fw and rw capabilities.*

| | FW-UAVs | RW-UAVs |
|---|---|---|
| Speed | Up to 100 m/s | Up to 30 m/s |
| Static Hovering | No | Yes |
| Altitude | Med-Low | Low |
| Movement degree | Low | High |

## Ground Base Station

A GBS, which is comprised of a transceiver, has the capability of broadcasting or receiving data recorder flow (including velocity, elevation, and battery status). A GBS is capable of interacting with any unmanned aerial vehicles (UAVs) in reach while also determining the reliability of their data transmission. Among some other things, the GBS is in responsibility of transmitting commands, such as planned heights, operating levels, and suitable velocity, to the appropriate aircraft (Hayat, S., et.al., 2016).

## User Interface

A radio remote controller is often used to control the movements of UAVs. Computers that are directly linked to GBSs are used to build and execute user interfaces. They let users to specify the activities to be completed as well as particular attributes like as update intervals. Furthermore, users have the flexibility to change the duties as required based on the scenario or events that have transpired in the region (Krichen, L., et.al., 2018).

## Propagation Model

Radiation properties are critical for the creation of any communication system since they determine how well a system communicates. FANETs have their own set of features, such as mobility, ground reflections affect, and a high likelihood of Communication via a line of vision (Sightline) among unmanned aircraft (UAVs), which allows for the mathematical modelling of each particular channel in each condition. With specific attention on U2U channels and U2G channels, it has been proposed a set of channel estimate methods for each kind of communication. The high-end computation is achieved through map reduce. (Srinivasakumar, V. et.al.,2022)

## Software Defined Networks (SDN)

SDN play a major role in monitoring the networks. It helps in maintaining the configuration changes whenever needed without any delay. It works independent of the vendor and scalability is achieved (Vanamoorthy, M. et.al.,2020)

## Band of Frequency

The overwhelming bulk of unmanned air vehicle (UAV) connectivity are endorsed by unlicensed spectrum like 0.9 GHz and 2.4 GHz, that are considered as inappropriate due to the possibility that they will become overcrowded with some other transceivers in the not-too-distant future. It has been proved that U2G lines may be implemented most effectively at the 5 GHz frequency. Furthermore, in order to prevent interfering with other frequencies, the 5.9 GHz frequency is believed to be the most appropriate, particularly when utilized in conjunction with IEEE 802.11p (Arnosti, S. Z., et.al.,2017).

## Energy Autonomy

One of the most significant difficulties with FANETs is energy usage. This is due to the fact that UAVs are powered by integrated batteries with limited energy capacity. Because the energy input needed for Uavs burning is substantially more than the amount of energy necessary for traditional aero plane burning. As a result, the development of communications networks should take into account consumption efficiency in order to extend the network's lifespan and reduce UAV failures (Dai, R., et.al.,2018).

## Localization

FANET needs precise localization with very brief amount of time update intervals owing to the fact that unmanned aerial vehicles (UAVs) fly at high velocities and have unexpected movement in some scenarios. Because location updates are available at one-second intervals, GPS is deemed inappropriate for A small number of FANET interfaces (e.g., commodity distribution, obstacle detection, building, and others) have been developed. Two types of localization algorithms are offered for these purposes: I Service orientation which is dependent on the interchange of packets, is another option. Maximum height localization, which is dependent on the elevations of unmanned aerial vehicles (UAVs).

## COVERAGE

UAVs are a cost-effective technique to represent a wide area, including in categorization or environmental control. situations. Furthermore, when terrestrial infrastructures are disrupted, UAVs may provide temporal connection coverage to ground users. UAVs use a variety of positioning approaches to provide good coverage, depending on the type of applications for which they are used (Kaleem, Z., et.al.,2018).

## Comparative Analysis

MANETs are classified into five separate networks, as shown in Figure 3, based on the kind of nodes and the contexts in which they are deployed. Each MANET sub-category and its functions are defined in this section.

*Figure 2. Manetand its sub-classes.*

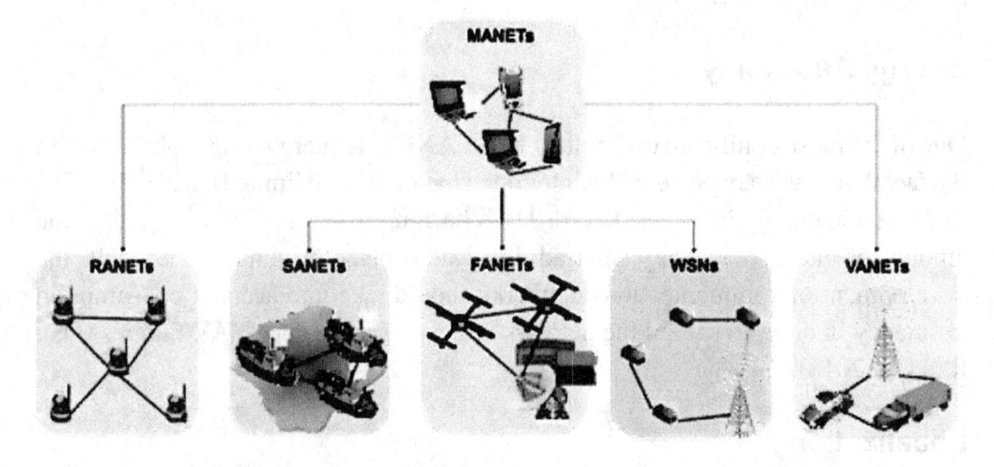


## Wireless Sensor Networks (WSNs)

Tiny and reduced devices constitute the foundation of these data-centric networks. Last but not least, there are sensors, which can both receive data from the environment and instantly transmit it to a centralized system known as a Sinks. Devices have many strategies have been suggested in the previous research to effectively regulate energy usage in order to extend the lifespan of WSNs (Marconato, E. A et.al.,2016).

### Vehicular Ad-hoc Networks (VANETs)

The mobile nodes of the systems in issue are mobile vehicles, making them a particular kind of MANET. The computing and resources capabilities of these devices is almost infinite. that is foreseeable and constrained by traffic patterns. Vehicles and road side units (RSUs) installed beside highways may communicate wirelessly with one other and with each other (Bacco, M., et.al.,2017).

### Robot Adhoc Networks (RANETs)

Because robots may be made up of transponders, and without centralized system, a Robot Ad hoc Net (RANET) is likely to be established as a mobile ad hoc network. In general, robot mobility may be smartly regulated to preserve network connection while providing a high packet transmission ratio. Robots' energy capacity is limited, necessitating careful monitoring of their power use.

### Ship Adhoc Networks (Sanets)

The major purpose of these huge network architectures is to increase maritime connectivity among boats. Due to their broad variety of uses, SANETs suffer the most from signal delay, which disrupts synchronized. Cross synchronizing mechanisms must be devised with the purpose of facilitating the interaction of massive ships with their coastal habitats (Vanitha, N., et.al.,2018).

## MOVEMENT DESIGNS FOR FANETS

The simulation side of things is where mobility models come in. These models are capable of simulating UAV behavior's in a realistic manner, allowing for the production of results that are as close to reality as feasible before a real-world deployment and test.

### Movement Model Types on Variational Data

Because of its ease, this type of networks has been widely used to characterize the motions of unmanned aerial vehicles (UAVs) and to monitor the success of FANETs. Indeed, each unmanned aerial vehicle (UAV) chooses its own movements at random, fully independent of the actions of other unmanned aerial vehicles (Li, Y., et.al.,2017).

RW (Random Walk) is a random-based model that enables mobile nodes to choose a random direction, speed, and distance at each fixed period of time t. The model is built on the concept of randomness. Generally speaking, as seen in Figure 5(a), RW exhibits abrupt shifts in direction. If you change directions, the new chosen direction becomes unrelated to the existing direction while you are changing directions. Several FANET protocols and applications, such as those described in have embraced this approach (Wu, H., et.al.,2018).

RWP (Random Way Point) is based on the same premise as RW but it adds wait durations in between directional changes between every change. As seen in Figure 5(b), in contrast to RW, the access points are more likely to occur in the center of the region of interest. RWP is used in a number of procedures involving unmanned aerial vehicles (UAVs) loitering at the same heights.

RD (Random Direction) is a technique meant to solve intensity fluctuations created by quasi-adjacent distribution induced by RWP particularly towards the center of the simulation area, by distributing the simulation area in a random manner. RD operates on the same premise as RW, in which mobile nodes choose a random direction, speed, and length to go in. As seen in Figure 5(c), the only variation between this and RW is that the node' dissemination is equal, irrespective of where they started off in the network. In the RD model is sorely tested in FANETs (Chakraborty, A., et.al.,2018).

*Figure 3. Designs randomized movement. (a) rwp. (b) rd. (c) mg.*

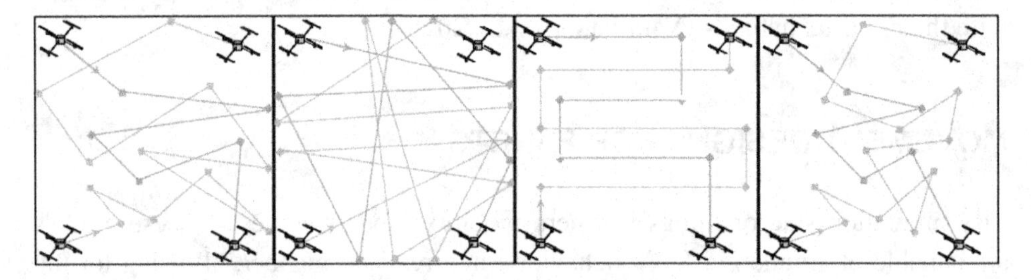

## Time-Based Mobility Models

A range of equations govern the movement of UAVs in this category., the present moment, and prior directions and speeds. Several of these factors are taken into account to ensure that motions are updated smoothly and that abrupt and harsh changes in direction and velocity are avoided.

There is a modelling approach known as LSA (Limitless Simulator Area), that only works in a small area. BSA attempts to circumvent this by turning the 2D rectangle simulation into a limitless torus-shaped simulation (see Figure 6(a)). In FANETs, this paradigm is not extensively used (Zeng, Y., et.al.,2016).

### GM(Gauss-Markov)

moment modelling approach that avoids abrupt mobility adjustments and can be tuned to varying degrees of unpredictability with only one scaling factor. Every network device has a set position and velocity at the start, as illustrated in Figure 6(b). Then, based on its prior velocity and direction, its future travel is determined. As a result of the effect of prior directions and speeds, GM can eliminate abrupt movement shifts and pauses.

### Improved EGM

It is the methodology that is devoted only to FANETs. The innovation here is in calculating the orientations of unmanned aerial vehicles (UAVs). Each UAV is given a random velocity and distance from a uniformly distributed wide speed range m/s and directions m/s, respectively (see Figure 6(c)). Several FANET applications make use of E-GM. (Mozaffari, M., et.al.,2019).

### ST (Smooth Turn)

FANET surveillance apps are supported by this device. ST provides for the recording of UAV trends and the creation of predictable paths (Example: a linear path or conventional large-radius twists. Figure 6(d) depicts an unmanned aerial vehicle (UAV) which hovers in circles around with a collection of randomly selected locations on a line parallel to its path of travel. This concept has been used by a number of services and applications.

## Path-Based Mobility Models

Each unmanned aerial vehicle (UAV) in this class is required to follow a specified course that has been pre-created and then loaded into the vehicle. When this predefined route is complete, the UAV may change direction and repeat the process (Orsino, A., et.al.,2017).

### CSM SR (Semi-Random Circular Movement)

UAVs are restricted to moving about a single point at the center with various circumference. After completing a complete turn, the UAV chooses a radius at random and moves at about the same static center again. SRCM has been utilized in a variety of projects. Paparazzi's movement concept, known as the PPMC, is a route modelling approach that employs five different motions: (c.f., Figure 4(b)). I Eighth, (1) Remain; 2) Scanning; 3) Oblong; 4) Waypoint; each conceivable movement a UAV may make is represented by a state machine in PPRZM. Many FANET protocols have utilised this paradigm.

FP (Flight Plan mobility model) creates a Functional Structure Topology (TDNT) map from a flight plan specified inside a document for movement (c.f., Figure 4(c)). FP is often utilised for if the entire flight path is planned out advance, air freight are two examples of where it has been used. (Khuwaja, A. A., et.al.,2018).

Because FANETs may function in heterogeneous networks, the MT (Multi-Tier Mobility Model) enables for different mobility patterns to be supported for example Airways and railroad systems, MT is a mixed modelling approach in which at least two distinct types of motions for various types of networks may be used (see Figure 4(d)). both utilise MT.

## Group-Based Mobility Models

UAVs prefer to travel together inside a designated zone designated by a point of reference to complete a task employing in a reasonable timeframe, FANETs. This causes UAVs to be spatially and temporally dependent on one another.

This is accomplished by forecasting the group's new positions in the following Figure 8(a) shows a specific time. Using ECR and FANETs, a cluster of Uavs can be controlled and avoided collisions. (Motlagh, N. H., et.al.,2016)

*Figure 4. Cluster multihop trajectory.*

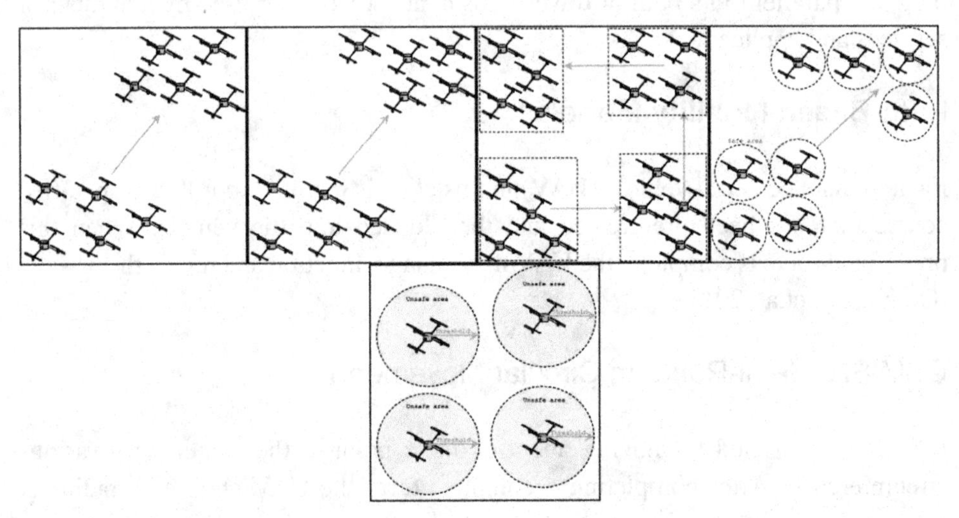

## PRS (Purse Mobility Model)

It is focused on a false system. The collection of movable nodes in PRS travels collectively to capture a certain target. The long-term location of the mobile node is calculated using a single updating formula that combines an accelerating component and a randomized vector, as shown in Figure 4 (c). In order to ensure effective monitoring of the randomized activity of each movable node is constrained by the objective being sought. PRS may be used whenever a team of UAVs tracks a stolen vehicle through with an urban area.

### PSMM (Particle Swarm Mobility Model)

UAV locations are calculated using a frame of reference. PSMM can predict each UAV's future velocity and distance based on its past performance

### STGM (Spatio Temporally Correlated Group Mobility Model)

It's a cluster analysis technique based on Gaussian Random type. STGM is based on the geographical correlations and also the temporal characteristics of simultaneously controlled UAVs' paths. (Hinzmann, T., et.al.,2016).

## Topology-Based Mobility Models

Operational connection limitation must be met indefinitely throughout time, actual power of UAV mobility is necessary. UAVs are necessary to be knowledgeable of their whole geometry in order to do so by coordinate their whereabouts with one another. When a network's connection must be maintained indefinitely, for example, UAV movements must be continually managed while preventing unnecessary erratic movements.

### SPR (Scattered Pheromone Repulsion)

SPR is a FANET-specific mobility model that use pheromones and localized search to direct UAVs to zones that have not been visited by other UAVs lately. A flying UAV drops virtual pheromones on the zone it has visited, which fade away with time. H3MP (Hybrid Stochastic Model with Pheromones) is a hybrid of DPR and Markov models that takes use of the benefits of both in a zone-decomposed environment. Using their prior positions in the inter-zone, the Markov model improves the overall motion of UAVs.

SDPC (Self-Deployable Point Coverage) is a FANET-specific topology-based mobility paradigm. SDPC may be used to improve the range of a large number of mobile nodes on the land while keeping UAVs connected. SDPC may utilize this model to deploy unmanned aerial vehicles (UAVs) over a disaster region to develop replacement infrastructure for the impacted population. (Van der Bergh, B., et.al.,2016).

## FANET ROUTING PROTOCOLS

One of the most difficult difficulties for FANETs is the design of the network level. This increases the pressure on researchers to come up with or adapt new routing protocols while meeting competing design constraints such as highly dynamic configuration balanced energy usage link fragmentation recovery, expandability security and wise utilize of the both UAV assets as well as apportioned frequency band. However, meeting all of the aforementioned criteria at the same time is very difficult, resulting in the classification of FANET routing based on the network's circumstances.

### Topology-Based Routing Protocols

There are a number of routing algorithms in this class that have been created initially for Mobile ad hoc networks. but have since been revised to account for the particular features of FANETs. These protocols use IP's of network nodes to transmit packets between communication nodes and are dependent on connection information. In this class, there are four subcategories: I Assertive, II react, III hybrid, and IV static. I proactive is the first category, II reactive is the second category, and so on. (Bittar, A., et.al.,2013).

### Assertive

In the routing tables, this category maintains information about all new linkages between every pair of mobile nodes. However, Because the design of FANETs is so variable, proactive routing protocols exchange a significant number of packets, causing the system to become congested and requiring a long time to react to shutdowns. If, and only if, some crucial upgrades are deployed, proactive steps may be warranted

## OLSR (Optimized Link State Routing Protocol)

OLSR (Optimized Link State Routing Protocol) has indeed been investigated in a number of recent research which have used OLSR in FANETs in a variety of simulated contexts. To reduce overhead, OLSR uses MultiPoint Relays (MPR) UAVs that cover two-hop neighbors, produce link status information, and forward packet forwarding to other MPRs. (Hong, Y., et.al.,2008)

## D-OLSR (Directional Optimized Link State Routing Protocol)

D-OLSR (Directional Optimized Link State Routing Protocol) is a variation of OLSR in which UAVs are outfitted with antenna arrays to extend transmission range. The amount of MPR UAVs has been drastically decreased to further reducing the gap. The distance between the source and destination UAVs is calculated by the source UAV. If the distance is more than Dmax/2 (i.e., Dmax is the distance supplied by a transmitting antenna), the MPR is chosen from the furthest UAV. D-OLSR, on the other hand, will shift to the traditional OLSR if the distance is less than Dmax/2. (Lu, J., et.al., 2018).

## ML-OLSR (Mobile and Load-aware Optimized Link State Routing Protocol)

To prevent choosing high-speed UAVs as MPRs, it is advocated. In ML-OLSR, the geographical placements and speeds of the neighbors are taken into account. Indeed, before each MPR selection, two metrics termed the Stabilization Degrees of Network (SDN) as well as the Reachability Level of Node (RDN) are computed (RDN). The MPRs are chosen depending on the SDN of each UAV in the vicinity. The chosen MPRs are thought to be likely to remain in range for an extended length of time in order to relay packets of data between the input and output UAVs. (Gong, J. et.al., 2018).

## DSDV (Destination-Sequenced Distance Vector)

Both to minimize network congestion and to update local knowledge on any topology change, two types of metrics are employed. In the scenario each UAV routing table and updated it with information about the whole network on a routine basis. Because higher part often, DSDV adds routing information to understand the most current routing route.

In comparison to OLSR TBRPF may give reduced latency in FANET, as shown. TBRPF has also been used in a set of FANET tests in which the connection strength of each route and the minimal hop count are taken into account. The UAVs create a route (i.e., a source tree) to all accessible UAVs in the system to get a global understanding of the networks. (Lu, J., et.al., 2018).

## REACTIVE

A route request process, also known as On-Demand routing algorithms, is started only if a UAV wishes to facilitate communications, and it explores, defines, and maintains as many routing pathways as possible. Owing to the discovery process, this group comes with a high delays and time lag in the majority of situations, as well as a substantial more than, particularly if the system is very scattered in nature DSR (Dynamic Source Routing) is a MANET-specific reactive routing system.

DSR's flexibility of loop property gives it the ability to choose from a variety of paths to any number of destination host. A number of papers in the literature, have used DSR to FANETs, and the bulk of them illustrate its weight and inability to cope with a large number of connection issues. In the instance shown in Figure 5(a), UAV S creates an RREQ packet and broadcast it to all of the other UAVs in the network to which it is connected. (Trotta, A. et.al., 2018).

1. The AODV (Ad hoc On-Demand Distance Vector) protocol is a combination that obtains capabilities such as hop-by-hop forwarding and periodical topologies table updates out of each of them and when used in combination with DSR and Variable sources protocol.

*Figure 5. Mechanism of reactive routing protocols. (a)Mechanism of DSR. (b) Mechanism of AODV. (c)Mechanism of TS-AODV. (d)Mechanism of M-AODV.*

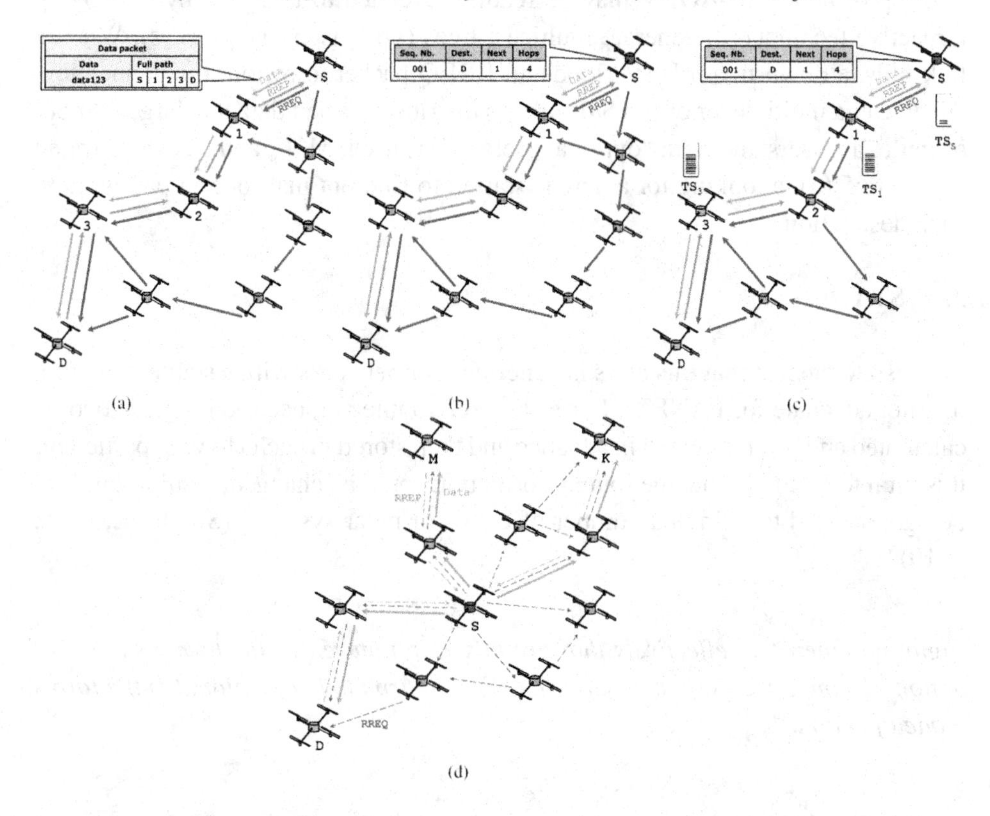


The novelty of the approaches is raised depending on the expiry time to preserve the constructed pathways, and the subsequent UAVs also change their packets. Many efforts to use AODV in FANETs have been made, including. When UAV S wants to interact with UAV D, it starts a route discovery process by distributing RREQ packets around the network to find the shortest (i.e., fewest hops) routing path (c.f., Figure 5 (b)).

The concept of scheduled slots is used in the TS-AODV (Flow Ad hoc On-Demand Length Vectors) protocol, that is a variant of the AODV. To decrease traffic concerns in FANETs, TS-AODV identified a trade-off between collision risk & bandwidth consumption (i.e., networks with a high number of UAVs). The length of each time slot allotted to each kind of packet is calculated based on a number of parameters, including the network's size, link fault diagnosis, and topology changes. Figure 5 depicts the operation of the TS-AODV (c). (Wang, Y., et.al., 2018).

2. M-AODV (Multicast Adhoc On-demand Distance Vector)

AODV has been updated to include the multipath routing idea for connecting a group of nodes. M-AODV may be readily extended to FANETs by employing a reactive technique to generate multicast trees (i.e., discovery process). When a link fails, a downstream UAV floods an RREQ packet to restore the connection. Whenever a multiplexer origin S desires to distribute packet data to a large number of multicast users, the term "optional protocol" is used. M, K, and D, as illustrated in Figure 5 (d), a looking for a new is started to find optimal routing pathways to each destination.

## 3. STATIC

Despite the fact that this class is beneficial for networks with a stable structure, it is not adequate for FANETs. For each UAV to interact, each routing protocol is calculated and pre-populated in advance and then stored in each UAV. In particular, it is vital to highlight that the route information could be changed, resulting in Uav being restricted to a limited number of UAVs or radar systems. (Singh, K., et.al., 2019).

*Figure 6. Operating effectively that are mixed in nature. (a) the hwmp's mode of action. (b) the zrp's exact mechanism. (c) the sharp mode of action. (d) the tora's mode of action.*

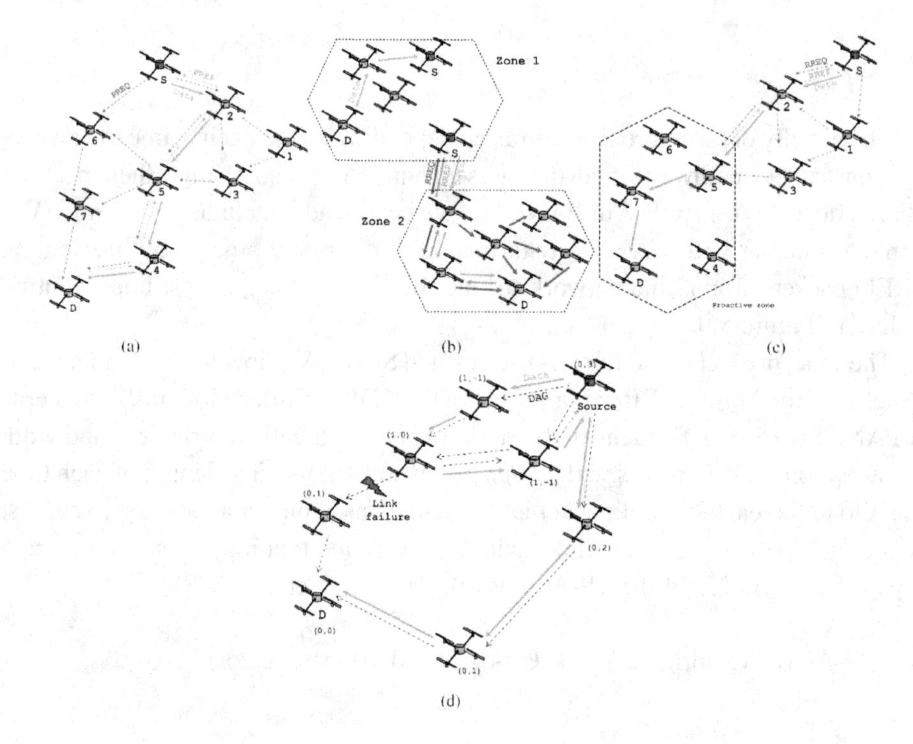

A common routing mechanism known as IHR (Inter Horizon Router) is used to solve the issue of adaptability that exists in a system. In order to do this, FANETs are organized into clusters, for each group possessing a cluster-head (CH) who symbolizes the entire group. Using this sort of routing in FANETs may be acceptable if the movement of UAVs is which was before by the presence of clusters or a huge number of Uav in a large network. In accordance with Figure 7(a), a CH, who is entirely responsible for interacting with the other CHs, the ground control station, and the fly nodes inside the cluster in which it is located supervise every clustering of flying terminals (Zheng, Z. et.al., 2018).

LCAD (Load Carry and Deliver) is a specialized static routing protocol for FANETs. Before UAVs take flight, LCAD configures the navigation route on the ground. By gathering datagrams, transporting them, and transmitting them to the specified destination, UAVs are regarded linkages here between pair of sender and receiver ground stations.

When a datagram is received by a variety of UAVs as well as the dissemination is done using a reactive mechanism, this may be done. DCR is built on a publish-subscribe approach, as illustrated in Figure 7(c), which may automatically link data producers to data subscribers. (Singh, K., et.al., 2019).

*Figure 7. Routing that are static in nature. (a) the fanet networking concept is based on the multi hierarchy. (a) the lcad action mechanism. (b) the dcr's mode of action.*

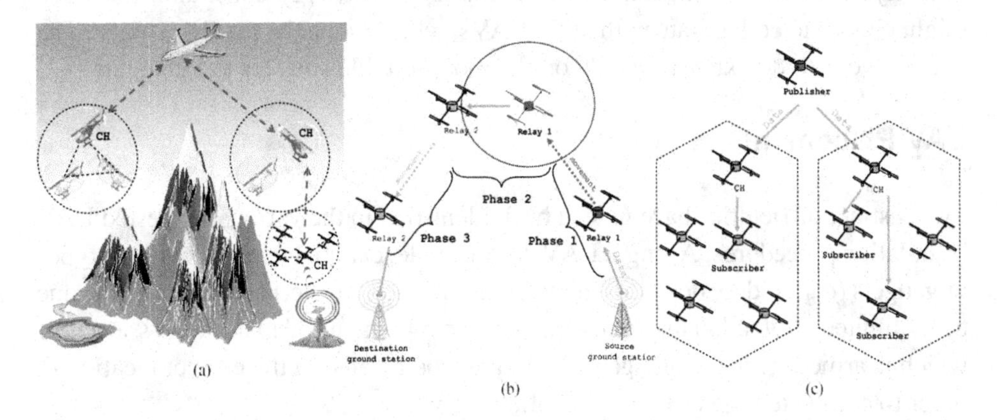


## OUTLOOKS ON DIFFICULTIES

Furthermore, this technology has the potential to enhance present services and offer new applications that have never been envisioned before. However, this technology confronts several challenges, including frequent disconnections, limited bandwidth, greater packet delay, and UAVs' low battery capacity.

## P2P UAV Communications

A swarm of UAVs needs peer-to-peer (P2P) connections for cooperative synchronization and obstacle detection. UAVs are suitable for P2P information and file transfer since they may operate as datasets. As a result, establishing novel P2P techniques and consolidating cast traffic for FANET connections might be a difficult task.

## Rules and Restrictions Regarding Commercial UAVS

Significant use in a variety of scenarios in our everyday lives, UAVs and their progress are incompatible with most nations' present airspace rules, which is their greatest roadblock. As a result, different laws for the use of UAVs in both commercial and scientific purposes to improve people's safety and confidentiality are urgently needed.

## Performance of the Energy Production

When powered by batteries, unmanned aerial vehicles (UAVs) have a limited power capacity, as is widely known. To overcome its own energy constraint, the UAV might choose to collaborate with other UAVs, which would be the first choice. The alternative is to do extensive study on the most suitable sites for fast chargers.

## UAV Placement

Two distinct difficulties have indeed been identified in the various suggested UAV installation procedures. A single UAV, for example, cannot be loaded with all types of gadgets (e.g., video, detectors, aircraft ground station, etc.) at the same time due to its limited payload. This necessitates the deployment of many UAVs, each of which is armed with a single gadget and must be located in the correct location in order to complete a specific mission effectively.

Furthermore, the energy usage of unmanned aerial vehicles (UAVs) is a big problem. Indeed, UAVs' residual energy is capable of not only supporting communications technology with around entities, but also supplying their many integrated components as well as motor energy. As a result, depending on the type of work, it is critical to prolong the UAV's operational lifespan. In this regard, optimising UAV deployment to determine the optimum needed number of UAVs as well as their locations remains an open question and a popular subject to investigate.

## COORDINATION OF UAVS WITH MANNED AIRCRAFT

Coordination among autonomously UAVs and other human aircraft is critical on many levels, including boosting zone communication, assuring the identification and destruction of questionable aircraft, and preventing collisions. As a consequence, a number of open challenge subjects have indeed been recognized and are prepared to be investigated, including regulating operations involving two different kinds of aircraft units, using a consistent band for communications, and increasing broadband service.

### Vulnerabilities Against Attacks

The level of protection provided by a network depends on how well it is secured. Because a network like FANET is not immune to hostile UAVs, its security is a major concern. Furthermore, because the data link layer is regarded to be of fundamental importance in supporting the most of services, it is necessary to examine this key problem.

### Crypto Algorithms and Secret Creation for Encryption Methods

Because UAVs communicate a lot of sensitive data, a hostile actor intercepting it would have severe consequences for this network. However, crypto security is exacerbated by the fact that unmanned aerial vehicles (UAVs) have a small battery and unable to execute crypto calculations efficiently. As a result, creating multiple security keys always has been a difficult task.

### Wireless Transmission Technology

UAVs have been able to communicate with one another by using existing wire-free broadcast frequencies (e.g., UHF and L-Band) that are already in use through different telecommunication systems, such as GSM and satellite links. Wireless LAN (WLAN) nodes have just been installed in unmanned aerial vehicles (UAVs) to enable for devices simultaneously in both the 2.4 GHz and 5 GHz bands, enabling for both U2U and Band interactions to take place.

### Establishing a Responsible Routing Protocol

Unmanned aerial vehicles (UAVs) in a FANET continually transfer packets of data with each other. making them very reliant on neighboring UAVs to complete a task. Which is why the communication As a result, developing a trustworthy network model is beneficial in reducing the vulnerability of non-trusted UAVs interacting with trustworthy UAVs is a key topic to investigate, and it is currently under investigation.

## Radio Propagation Models

Any interaction between a set of sending and receiving UAVs uses a propagation model. This model must account for various factors, including velocity, elevation, and acceleration angles, which determine the radio channel's properties. This is a topical and hot issue that is now in the conceptual phase, needing a significant amount of research and analysis in order to develop an effective modelling approach for such forms of communication channels

## The Aspect of Wireless Sensor Networks

Only a few studies incorporating WSN and UAV communication have been completed. Indeed, UAVs may be used as a cellular data collecting for sensors on the surface owing to high and regulated mobility, while keeping in mind their energy limitation capacity. As a result, WSNs are a popular subject with various difficulties to examine, including UAV movement and altitude.

## Uavs Control Through Cloud-Based System

Due to its limited energy supplies and computation capacity of UAV, an appropriate solution must be implemented to fully use all of the UAVs' available resources. A recent study recommended integrating the uses cloud paradigm with FANETs to overcome these difficulties. Cloud computing is a highly popular platform today, and it may be regarded an excellent way to increase the capacity of such networks' resources. Furthermore, cloud-based monitoring and management systems are among the hot subjects that are only nominally explored and have yet to be implemented. (Wang, H., et.al., 2018).

## Routing at the Cross-Layer Level

The majority of the AODV protocol covered in this article are solely concerned with concerns with a single network level (the Network layer), that is in charge of maintaining communication among UAVs. Cross-layer techniques may give additional flexibility by allowing all levels to share information about a specific network scenario by developing new connections between them and then reacting appropriately. The topic of forwarding cross-layer communications in FANETs has received little attention and is currently a work in progress.

## Routing Challenges

The rapid mobility and low population density of UAVs are important challenges in developing an effective routing scheme that ensures reliable data flow between them. When UAVs travel in 3D space, the intensity of these challenges increases (i.e., different altitudes). Based on a particular circumstance found in the network, many techniques have been proposed throughout the literature. As a result, there is a pressing need to develop new protocols that can use the proper approach in a specific circumstance. (Khan, M. A., et.al., 2018).

## CONCLUSION

Routing is one of the most important components of FANETs to guarantee proper functioning and effective cooperative entire network. Approximately sixty methods have been suggested in the research over the past decade. Each has its own traits, characteristics, disadvantages, and key success factors. This thorough study assessed the most essential aspects which have a strong association with FANET routing to distinguish them & put it into perspective. In order to demonstrate the novelty of our survey, we first selected the bulk of the questionnaire works and evaluated them subjectively based on outstanding features. The study then defines the design of FANETs in an innovative approach by characterizing the many accepted organizations, the current kinds of communication, and their features. A quick comparative study of Mobile ad - hoc post in term of numerous key aspects has also been provided so that A thorough awareness of the most challenging kind of network exists among its users. Lastly, since modeling approaches are so critical in assessing the success of a routing algorithm, we examined the Recent FANET mobility simulations have been categorized per a classification method, and these versions are listed below .and then evaluated using various criteria.

The most often used strategies by the Distributed network routing algorithms are explained in the second stage. Following that, a worldwide classification of the Data forwarding routing is offered, which divides the protocols into eight primary groups and 10 subgroups. Every subcategory is explained individually, with explanatory figures for its routing methods, which are then contrasted depending on several features.

Finally, we highlighted the open research issues and needs for FANET routing that have received less attention. Furthermore, for researchers who want to go further into this study topic, we've supplied some viable answers as well as some suggested resources. As a definitive decision to this research, Accordingly, FANET route solutions must be able to deal the networking splitting and the show's very dynamic structure. Future views that we are now researching include specializing in the UAV-assisted idea, which has received less attention in the past but has lately piqued the curiosity of a large number of scientists. Furthermore, we want to devise a route optimization protocol, which can be tailored to any case while taking into account the many restrictions explored.

## REFERENCES

Altawy, R., & Youssef, A. M. (2016). Security, privacy, and safety aspects of civilian drones: A survey. *ACM Transactions on Cyber-Physical Systems*, *1*(2), 1–25. doi:10.1145/3001836

Arafat, M. Y., & Moh, S. (2018). A survey on cluster-based routing protocols for unmanned aerial vehicle networks. *IEEE Access: Practical Innovations, Open Solutions*, *7*, 498–516. doi:10.1109/ACCESS.2018.2885539

Arnosti, S. Z., Pires, R. M., & Branco, K. R. (2017, June). Evaluation of cryptography applied to broadcast storm mitigation algorithms in FANETs. In *2017 International Conference on Unmanned Aircraft Systems (ICUAS)* (pp. 1368-1377). IEEE. 10.1109/ICUAS.2017.7991377

Bacco, M., Cassará, P., Colucci, M., Gotta, A., Marchese, M., & Patrone, F. (2017, September). A survey on network architectures and applications for nanosat and UAV swarms. In *International Conference on Wireless and Satellite Systems* (pp. 75-85). Springer.

Bittar, A., & de Oliveira, N. M. (2013). Central processing unit for an autopilot: Description and hardware-in-the-loop simulation. *Journal of Intelligent & Robotic Systems*, *70*(1), 557–574. doi:10.100710846-012-9745-y

Campion, M., Ranganathan, P., & Faruque, S. (2018, May). A review and future directions of UAV swarm communication architectures. In *2018 IEEE international conference on electro/information technology (EIT)* (pp. 903-908). IEEE.

Chakraborty, A., Chai, E., Sundaresan, K., Khojastepour, A., & Rangarajan, S. (2018, December). SkyRAN: a self-organizing LTE RAN in the sky. In *Proceedings of the 14th International Conference on emerging Networking EXperiments and Technologies* (pp. 280-292). 10.1145/3281411.3281437

Cumino, P., Lobato, W. Junior, Tavares, T., Santos, H., Rosário, D., Cerqueira, E., ... Gerla, M. (2018). Cooperative UAV scheme for enhancing video transmission and global network energy efficiency. *Sensors (Basel)*, *18*(12), 4155. doi:10.339018124155 PMID:30486376

da Cruz, E. P. F. (2018). A comprehensive survey in towards to future FANETs. *IEEE Latin America Transactions*, *16*(3), 876–884. doi:10.1109/TLA.2018.8358668

Dai, R., Fotedar, S., Radmanesh, M., & Kumar, M. (2018). Quality-aware UAV coverage and path planning in geometrically complex environments. *Ad Hoc Networks*, *73*, 95–105. doi:10.1016/j.adhoc.2018.02.008

Fan, X., Cai, W., & Lin, J. (2017, October). A survey of routing protocols for highly dynamic mobile ad hoc networks. In *2017 IEEE 17th International Conference on Communication Technology (ICCT)* (pp. 1412-1417). IEEE. 10.1109/ICCT.2017.8359865

Gong, J., Chang, T. H., Shen, C., & Chen, X. (2018). Flight time minimization of UAV for data collection over wireless sensor networks. *IEEE Journal on Selected Areas in Communications*, *36*(9), 1942–1954. doi:10.1109/JSAC.2018.2864420

Hayat, S., Yanmaz, E., & Muzaffar, R. (2016). Survey on unmanned aerial vehicle networks for civil applications: A communications viewpoint. *IEEE Communications Surveys and Tutorials*, *18*(4), 2624–2661. doi:10.1109/COMST.2016.2560343

Hinzmann, T., Stastny, T., Conte, G., Doherty, P., Rudol, P., Wzorek, M., ... Gilitschenski, I. (2016, October). Collaborative 3d reconstruction using heterogeneous uavs: System and experiments. In *International Symposium on Experimental Robotics* (pp. 43-56). Springer.

Hong, Y., Fang, J., & Tao, Y. (2008, October). Ground control station development for autonomous UAV. In *International Conference on Intelligent Robotics and Applications* (pp. 36-44). Springer. 10.1007/978-3-540-88518-4_5

Jiang, J., & Han, G. (2018). Routing protocols for unmanned aerial vehicles. *IEEE Communications Magazine*, *56*(1), 58–63. doi:10.1109/MCOM.2017.1700326

Kaleem, Z., & Rehmani, M. H. (2018). Amateur drone monitoring: State-of-the-art architectures, key enabling technologies, and future research directions. *IEEE Wireless Communications*, *25*(2), 150–159. doi:10.1109/MWC.2018.1700152

Khan, M. A., Khan, I. U., Safi, A., & Quershi, I. M. (2018). Dynamic routing in flying ad-hoc networks using topology-based routing protocols. *Drones*, *2*(3), 27. doi:10.3390/drones2030027

Khan, M. A., Safi, A., Qureshi, I. M., & Khan, I. U. (2017, November). Flying ad-hoc networks (FANETs): A review of communication architectures, and routing protocols. In *2017 First international conference on latest trends in electrical engineering and computing technologies (INTELLECT)* (pp. 1-9). IEEE.

Khuwaja, A. A., Chen, Y., Zhao, N., Alouini, M. S., & Dobbins, P. (2018). A survey of channel modeling for UAV communications. *IEEE Communications Surveys and Tutorials*, *20*(4), 2804–2821. doi:10.1109/COMST.2018.2856587

Krichen, L., Fourati, M., & Fourati, L. C. (2018, September). Communication architecture for unmanned aerial vehicle system. In *International Conference on Ad-Hoc Networks and Wireless* (pp. 213-225). Springer. 10.1007/978-3-030-00247-3_20

Li, Y., & Cai, L. (2017). UAV-assisted dynamic coverage in a heterogeneous cellular system. *IEEE Network*, *31*(4), 56–61. doi:10.1109/MNET.2017.1600280

Lu, J., Wan, S., Chen, X., Chen, Z., Fan, P., & Letaief, K. B. (2018). Beyond empirical models: Pattern formation driven placement of UAV base stations. *IEEE Transactions on Wireless Communications*, *17*(6), 3641–3655. doi:10.1109/TWC.2018.2812167

Marconato, E. A., Maxa, J. A., Pigatto, D. F., Pinto, A. S., Larrieu, N., & Branco, K. R. C. (2016, June). IEEE 802.11 n vs. IEEE 802.15.4: A study on communication QoS to provide safe FANETs. In *2016 46th Annual IEEE/IFIP international conference on dependable systems and networks workshop (DSN-W)* (pp. 184-191). IEEE.

Maxa, J. A., Mahmoud, M. S. B., & Larrieu, N. (2017). Survey on UAANET routing protocols and network security challenges. *Ad-Hoc & Sensor Wireless Networks*, 37.

Motlagh, N. H., Taleb, T., & Arouk, O. (2016). Low-altitude unmanned aerial vehicles-based internet of things services: Comprehensive survey and future perspectives. *IEEE Internet of Things Journal*, *3*(6), 899–922. doi:10.1109/JIOT.2016.2612119

Mozaffari, M., Saad, W., Bennis, M., Nam, Y. H., & Debbah, M. (2019). A tutorial on UAVs for wireless networks: Applications, challenges, and open problems. *IEEE Communications Surveys and Tutorials*, *21*(3), 2334–2360. doi:10.1109/COMST.2019.2902862

Orsino, A., Ometov, A., Fodor, G., Moltchanov, D., Militano, L., Andreev, S., Yilmaz, O. N. C., Tirronen, T., Torsner, J., Araniti, G., Iera, A., Dohler, M., & Koucheryavy, Y. (2017). Effects of heterogeneous mobility on D2D-and drone-assisted mission-critical MTC in 5G. *IEEE Communications Magazine*, *55*(2), 79–87. doi:10.1109/MCOM.2017.1600443CM

Oubbati, O. S., Lakas, A., Zhou, F., G✓neş, M., & Yagoubi, M. B. (2017). A survey on position-based routing protocols for Flying Ad hoc Networks (FANETs). *Vehicular Communications, 10*, 29–56. doi:10.1016/j.vehcom.2017.10.003

Park, S. Y., Shin, C. S., Jeong, D., & Lee, H. (2018). DroneNetX: Network reconstruction through connectivity probing and relay deployment by multiple UAVs in ad hoc networks. *IEEE Transactions on Vehicular Technology, 67*(11), 11192–11207. doi:10.1109/TVT.2018.2870397

Sánchez-García, J., García-Campos, J. M., Arzamendia, M., Reina, D. G., Toral, S. L., & Gregor, D. (2018). A survey on unmanned aerial and aquatic vehicle multi-hop networks: Wireless communications, evaluation tools and applications. *Computer Communications, 119*, 43–65. doi:10.1016/j.comcom.2018.02.002

Sharma, V., & Kumar, R. (2017). Cooperative frameworks and network models for flying ad hoc networks: A survey. *Concurrency and Computation, 29*(4), e3931. doi:10.1002/cpe.3931

Singh, K., & Verma, A. K. (2019). Flying adhoc networks concept and challenges. In *Advanced methodologies and technologies in network architecture, mobile computing, and data analytics* (pp. 903–911). IGI Global.

Srinivasakumar, V., Vanamoorthy, M., Sairaj, S., & Ganesh, S. (2022). An alternative C++-based HPC system for Hadoop MapReduce. *Open Computer Science, 12*(1), 238–247. doi:10.1515/comp-2022-0246

Tareque, M. H., Hossain, M. S., & Atiquzzaman, M. (2015, September). On the routing in flying ad hoc networks. In *2015 federated conference on computer science and information systems (FedCSIS)* (pp. 1-9). IEEE.

Thammawichai, M., Baliyarasimhuni, S. P., Kerrigan, E. C., & Sousa, J. B. (2017). Optimizing communication and computation for multi-UAV information gathering applications. *IEEE Transactions on Aerospace and Electronic Systems, 54*(2), 601–615. doi:10.1109/TAES.2017.2761139

Trotta, A., Di Felice, M., Montori, F., Chowdhury, K. R., & Bononi, L. (2018). Joint coverage, connectivity, and charging strategies for distributed UAV networks. *IEEE Transactions on Robotics, 34*(4), 883–900. doi:10.1109/TRO.2018.2839087

Van der Bergh, B., Chiumento, A., & Pollin, S. (2016). LTE in the sky: Trading off propagation benefits with interference costs for aerial nodes. *IEEE Communications Magazine, 54*(5), 44–50. doi:10.1109/MCOM.2016.7470934

Vanamoorthy, M., & Chinnaiah, V. (2020). Congestion-free transient plane (CFTP) using bandwidth sharing during link failures in SDN. *The Computer Journal*, *63*(6), 832–843. doi:10.1093/comjnl/bxz137

Vanamoorthy, M., Chinnaiah, V., & Sekar, H. (2020). A hybrid approach for providing improved link connectivity in SDN. *The International Arab Journal of Information Technology*, *17*(2), 250–256. doi:10.34028/iajit/17/2/13

Vanitha, N., & Padmavathi, G. (2018, March). A comparative study on communication architecture of unmanned aerial vehicles and security analysis of false data dissemination attacks. In *2018 International Conference on Current Trends towards Converging Technologies (ICCTCT)* (pp. 1-8). IEEE. 10.1109/ICCTCT.2018.8550873

Wang, H., Ding, G., Gao, F., Chen, J., Wang, J., & Wang, L. (2018). Power control in UAV-supported ultra-dense networks: Communications, caching, and energy transfer. *IEEE Communications Magazine*, *56*(6), 28–34. doi:10.1109/MCOM.2018.1700431

Wang, J., Jiang, C., Han, Z., Ren, Y., Maunder, R. G., & Hanzo, L. (2017). Taking drones to the next level: Cooperative distributed unmanned-aerial-vehicular networks for small and mini drones. *IEEE Vehicular Technology Magazine*, *12*(3), 73–82. doi:10.1109/MVT.2016.2645481

Wang, Y., Sun, T., Rao, G., & Li, D. (2018). Formation tracking in sparse airborne networks. *IEEE Journal on Selected Areas in Communications*, *36*(9), 2000–2014. doi:10.1109/JSAC.2018.2864374

Wu, H., Tao, X., Zhang, N., & Shen, X. (2018). Cooperative UAV cluster-assisted terrestrial cellular networks for ubiquitous coverage. *IEEE Journal on Selected Areas in Communications*, *36*(9), 2045–2058. doi:10.1109/JSAC.2018.2864418

Zeng, Y., Zhang, R., & Lim, T. J. (2016). Wireless communications with unmanned aerial vehicles: Opportunities and challenges. *IEEE Communications Magazine*, *54*(5), 36–42. doi:10.1109/MCOM.2016.7470933

Zhao, H., Wang, H., Wu, W., & Wei, J. (2018). Deployment algorithms for UAV airborne networks toward on-demand coverage. *IEEE Journal on Selected Areas in Communications*, *36*(9), 2015–2031. doi:10.1109/JSAC.2018.2864376

Zheng, Z., Sangaiah, A. K., & Wang, T. (2018). Adaptive communication protocols in flying ad hoc network. *IEEE Communications Magazine*, *56*(1), 136–142. doi:10.1109/MCOM.2017.1700323

# Chapter 3
# Evolution of Vehicular Ad Hoc Network and Flying Ad Hoc Network for Real-Life Applications:
## Role of VANET and FANET

**Sipra Swain**
*National Institute of Technology, Rourkela, India*

**Biswa Ranjan Senapati**
*ITER, Siksha 'O' Anusandhan University (Deemed), India*

**Pabitra Mohan Khilar**
*National Institute of Technology, Rourkela, India*

## ABSTRACT

*The demand for the quick transmission of data at any point and at any location motivates researchers from the industry and academics to work for the enhancement of ad hoc networks. With time, various forms of ad hoc networks are evolved. These are MANET, VANET, FANET, AANET, WSN, SPAN, etc. The initial objective of VANET is to provide safety applications by combining them with ITS. But later, the applications of VANET are extended to commercial, convenience, entertainment, and productive applications. Similarly, connections among multiple unmanned aerial vehicles (UAV) through wireless links, architectural simplicity, autonomous behaviour of UAV, etc. motivate the researchers to use FANET in various sectors like military, agriculture, and transportation for numerous applications. Search and rescue operations, forest fire detection and monitoring, crop management monitoring, area mapping, and road traffic monitoring are some of the applications of FANET. The authors mentioned some applications in the chapter using VANET, FANET, and the combination of VANET and FANET.*

DOI: 10.4018/978-1-6684-3610-3.ch003

## INTRODUCTION

Advancement in wireless access technology, demands in communication at anytime and at any location, availability of various sensors at affordable cost increases the demand for the ad hoc network (Helen & Arivazhagan, 2014). As per the demand for communication, various forms of ad hoc networks are evolved (Al-Absi & Lee, 2021). Different ad hoc networks are mobile ad hoc networks (MANET), vehicular ad hoc networks (VANET), flying ad hoc networks (FANET), airlift ad hoc networks (AANET), wireless sensor networks (WSN), smartphone ad hoc networks (SPAN), etc. Overall comparison for the different forms of ad hoc networks is presented in Table 1.

*Table 1. Comparison of different forms of ad hoc networks*

| Characteristics | MANET | VANET | FANET | AANET | WSN | SPAN |
|---|---|---|---|---|---|---|
| Network connectivity | High | Medium | Low | Low | High | Medium |
| Availability of energy | Low | High | Low | Low | Low | Medium |
| Mobility models | Random | Restricted | Restricted | Random | Restricted | Restricted |
| Speed of the node | Medium | High | High | High | Low | Low |
| Link connectivity | Changes occasionally | Changes frequently | Changes frequently | Changes frequently | Changes occasionally | Changes Occasionally |

(Zhang, et al., 2019)

The vehicular ad hoc network (VANET) is one of the subclasses of ad hoc networks (Hamdi, Audah, Rashid, Mohammed, Alani, & Mustafa, A review of applications, characteristics and challenges in vehicular ad hoc networks (VANETs), 2020). VANET consists of two components which are distributed and asynchronous (Senapati B. R., 2021). These two components are as follows.

1. **Vehicle:** These nodes are mobile.
2. **Roadside unit (RSU):** These nodes are static.

From the above-mentioned two components, RSU has greater sensing capability, computational power, and greater storage capacity. As RSU is static in nature, RSU is generally placed near the junction. Technological advancement in the electro-mechanical-computational-humanity department, availability of a large number of sensors at affordable cost, presence of communication unit (On-board Unit-OBU), an increasing number of vehicular nodes for better network connectivity transforms the traditional carrier vehicle into smart and intelligent vehicles (Sun, 2017). Figure 1 shows various components of smart and intelligent vehicles. Various components of smart and intelligent vehicles are GPS, front and rear radar, OBU, in-vehicle sensors, computational platform, event-data recorded, etc. Along with VANET, Flying ad hoc network (FANET) is also one of the current research areas whose brief introduction is mentioned below.

*Figure 1. Components of smart and intelligent vehicle*

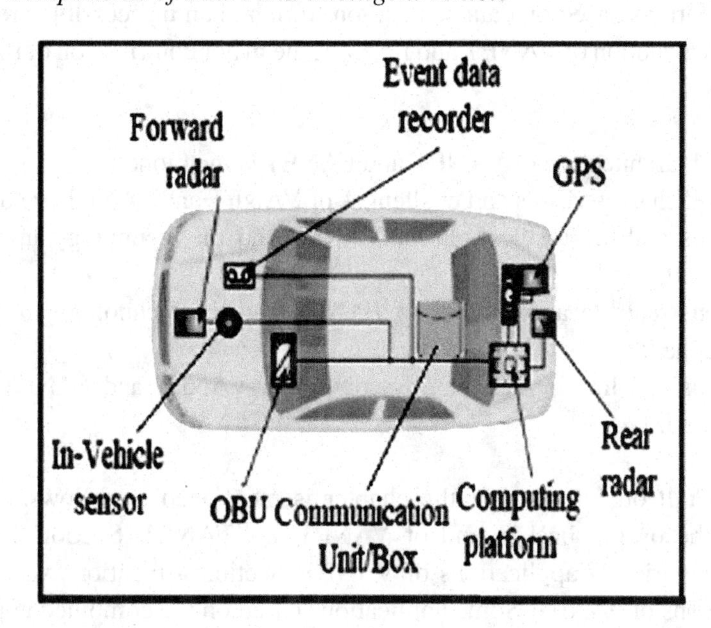


An unmanned aerial vehicle (UAV) is a flying vehicular system without any human pilot. It is a pre-programmed or remote-controlled system. In recent decades the progress in the field of robotics technology includes electronics, sensors, facilitation in the process of data transmission, make a feasible production of UAVs for various applications. The progressing nature of robotics technologies brings many advances in UAV design and production in terms of its flexibility and agility, payload, sensing capability, economical cost, etc. The above-mentioned characteristics of UAVs are the reasons behind their popularity in both military and civilian applications. The UAV-based application started with a single UAV system, where an individual large UAV is responsible for the accomplishment of the whole task. Single UAV systems are mostly used in surveillance and monitoring applications. However, the successive implication of various large-scale applications is difficult for a single

UAV system because of its restricted range of service and more vulnerability towards mission failure. Therefore, a collaborative work of a set of small UAVs grabs the whole attention these days (Singh, 2015). A group of homogeneous or heterogeneous UAVs collectively form an ad hoc network known as flying ad hoc network (FANET) (Chriki, Touati, Snoussi, & Kamoun, 2019), where a multi-UAV system works co-operatively for the successful execution of a mission (Srivastava & Prakash, Future FANET with application and enabling techniques: Anatomization and sustainability issues, 2021). The multi-UAV system increases the robustness of the network as well as the range of the service providence. FANETs are used in most applications based on data acquisition from human inaccessible areas. After the brief introduction of VANET and FANET, the major contribution of the chapter is as follows.

1. Overall architecture of VANET and FANET is mentioned.
2. Overall characteristics and challenges of VANET and FANET are discussed.
3. Various real-life applications of VANET and the technology involved are specified.
4. Various real-life applications of FANET and the technology involved are mentioned.
5. Various real-life applications by combining VANET and FANET are also specified.

The overall organization of the chapter is mentioned as follows. Section 2 discusses the overall background of VANET and FANET. Section 3 specifies various categories of applications of VANET. Section 4 mentions various types of applications of FANET. Some applications based on the combined approach of VANET and FANET are discussed in Section 5. Finally, the conclusion is mentioned in the last section i.e. in Section 6.

## BACKGROUND OF VANET AND FANET

This section discusses various components, architectures, characteristics, and challenges of VANET and FANET in real-life applications respectively. Various components of VANET are mentioned in the next sub-section.

### Components of VANET

Various components of VANET and their role in VANET (Goyal, Agarwal, & Tripathi, 2019)are mentioned as follows.

1. **Communication unit/box:** The role of the communication box is to provide the services like I-Call, B-Call, and E-Call with the help of a mobile network.
2. **Front radar and rear radar:** Front radar and rear radar play a crucial role in Advance Driver Assistance Systems (ADAS). ADAS includes the detection of static and dynamic objects around the car's vicinity, automatic emergency braking system (AEBS), and adaptive cruise control (ACC).
3. **Geo-positioning system (GPS):** The presence of GPS in the vehicle helps to determine the location of the vehicle, vehicular speed, time, and direction of motion of the vehicle.
4. **On-board unit (OBU):** It is the communication unit for the vehicle through which a vehicle can communicate with other vehicles or roadside units.
5. **In-vehicle sensors:** In-vehicle sensors make the vehicle intelligent by controlling different aspects such as temperature, coolant levels, oil pressure, emission levels, and much more. Different types of in-vehicle sensors present in a vehicle are vehicle speed sensor, mass air flow sensor, coolant sensor, fuel sensor, ultra-sonic sensor, etc.
6. **Event data recorder:** The event data recorder (EDR) records important vehicle data associated with the vehicle's accident.

## Architecture of VANET

The objective of the architecture of VANET is to provide effective communication among the vehicles moving on the road network, between vehicles and the equipment present statically on the roadside, and with the Internet infrastructure for authentication, security, etc (Ahmed, Pierre, & Quintero, 2017). The overall architecture of VANET is presented in Figure 2.

*Figure 2. Architecture of VANET*

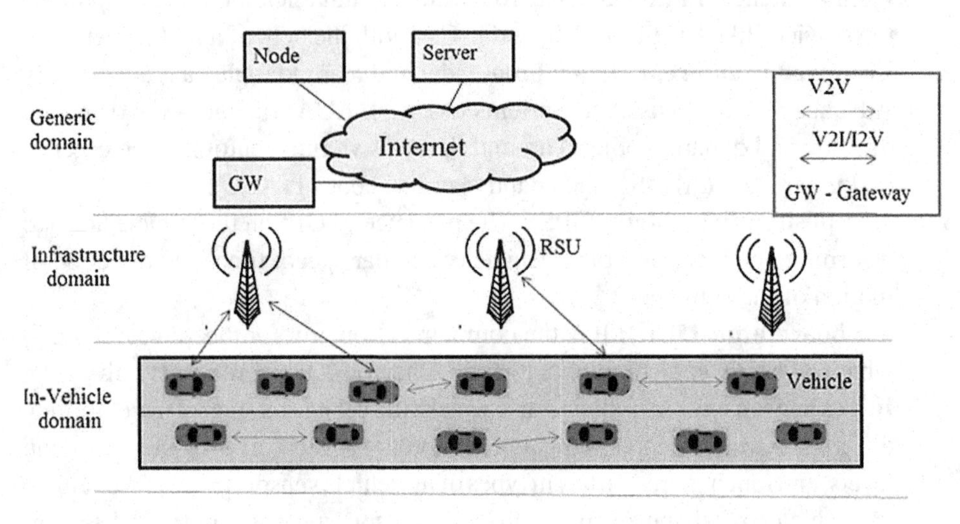


The overall architecture can be categorized into three domains. These are the in-vehicle domain, infrastructure domain, and generic domain (Liang, Li, Zhang, Wang, & Bie, 2015).

- **In-vehicle domain:** In-vehicle domain consists of mobile vehicles equipped with OBU and application units. The presence of OBU in the vehicle is responsible forV2V communication.
- **Infrastructure domain:** The infrastructure domain consists of static entities called RSU.RSU acts as an access point that supports V2I/I2V communication in VANET. The infrastructure domain also consists of the central managing centre like traffic management centre, vehicle management centre, etc.
- **Generic domain:** It consists of Internet infrastructure and Private infrastructure. Different components working for VANET like node, server, etc., comes under the category of the generic domain.

The next subsection focuses on the characteristics of VANET.

## Characteristics of VANET

Various characteristics of VANET are discussed as follows (Hamdi, Audah, Rashid, Mohammed, Alani, & Mustafa, A review of applications, characteristics and challenges in vehicular ad hoc networks (VANETs), 2020).

- A frequent and rapid change in network topology.
- High mobility of the vehicular nodes.
- Unbounded network size (can be implemented for one city, several cities, or for a country).
- Virtually no power constraint.
- Frequent data dissemination among the nodes of VANET.
- No computation resource and memory constraint.
- Geographic location-dependent communication.

Although the characteristics of VANET is unique and different from other wireless networks,

Some of the characteristics impose some challenges for the successful deployment of VANET. The next subsection discusses the challenges for the successful deployment of VANET.

## Challenges in VANET

Various challenges in VANET are as follows (Ur Rehman, Khan, Zia, & Zheng, 2013).

1. **Highly dynamic topology:** Characteristic of the radio propagation and the high speed of the vehicle in two-way directions are responsible for the dynamic topology of VANET. Vehicles can quickly join or leave the network soon, resulting in fast and frequent topology changes.
2. **Congestion and collision control:** The unbounded network size creates a challenge of network data congestion and data packet collision.
3. **Faulty vehicle:** The presence of a faulty component of a vehicle in VANET significantly reduces the performance of the VANET. The faulty component may be the in-vehicle sensors, GPS, or the vehicle's communication unit (OBU). The faulty part may affect the routing, reduce the packet delivery ratio, transmit incorrect data, etc.
4. **Security:** Since the information is transmitted through a wireless medium, so security is one of the important, challenging issues. During the propagation of the data packet, the information should not be modified by the attackers. The security aspect must consider authentication, confidentiality, non-repudiation, integrity, etc.
5. **Environmental impact:** For the communication between the nodes of VANET, electromagnetic waves are used. The propagation of these waves is affected by the environment. The presence of obstacles also affects the overall propagation.

The next subsection discusses the background of FANET which includes the components of FANET, the architecture of FANET, characteristics, and challenges of FANET.

## Components of FANET

A UAV consists of different modular components, which are incorporated within the body of the UAV. These components are responsible for the automated operation of UAVs as well as the successful accomplishment of various mission-specific requirements. The main components of UAV include various sensors, flight control unit (FCU), payload, and communication equipment (Alghamdi, Munir, & La, 2021; Adão, et al., 2017).

1. **Sensors:** UAV needs several sensors to gather data for different purposes such as obstacle avoidance, mission-oriented data acquisition, situational awareness, etc. Some important sensors used in UAV systems are inertial measurement unit (IMU), global positioning system (GPS), visual, infrared sensors, etc. UAV uses the GPS and IMU sensors together for determining the position, speed, and heading direction of itself. The navigation system of UAV used these above parameters for path planning. On the other hand, visual and infrared sensors are used for avoiding obstacles by sensing the environment.
2. **Flight control unit (FCU):** FCU is used for the autonomous navigation of the UAV system. It processes all the information gathered by IMU, GPS, and other sensors for the navigation assistance of UAVs.
3. **Payload:** The amount of extra weight a UAV can bring with itself is termed as a payload. It is usually measured other than its own weight that includes the additional weight due to various sensors and cameras, delivery items, various weapons in case of military applications, container, and sprayer for agricultural applications, etc.
4. **Communication equipment:** In UAV based application system, the communication between the UAV operator and the UAV can be established through any of the following: the ground control station (GCS), radio controller transmitter, or satellite. The UAV, which is connected with the ground controller, has two radio frequency data links for communication such as uplink and downlink. Uplink is used to transmit the control command from the ground controller to UAV; however, the downlink is used to carry the required information such as position, speed, direction, energy level, and other mission-specific information from UAV to the ground controller. On the other hand, the antenna is an essential part of the communication system of a UAV. The structure of the antenna plays an important role in the communication

process between UAVs. Two types of antennas are used for various FANET applications such as omnidirectional and directional antenna (Bekmezci, Sahingoz, & Temel., 2013). A comparison of both types of antennas is given in Table 2.

*Table 2. A comparison of omnidirectional and directional antenna attributes*

| Characteristics | Omnidirectional | Directional |
|---|---|---|
| Signal direction | Every direction | Desired |
| Node orientation | Not required | Required |
| Transmission range | Smaller | Larger |
| Vulnerable for Jamming | More susceptibility | Lesser susceptibility |

Based on the components, UAVs are classified into different categories such as micro UAV (µUAV), micro air vehicle (MAV), nano air vehicle (NAV), and pico air vehicle (PAV) (Hassanalian & Abdelkefi, 2017). Table 3 represents the details of the above classification.

*Table 3. Classification of UAV based on size and weight*

| Designation | Weight Range (W) | Wingspan Range ($W_s$) |
|---|---|---|
| UAV | 5 kg < W <= 15000 kg | $2\ \text{m} < W_s <= 61\ \text{m}$ |
| µUAV | 2 kg < W <= 5 kg | $1\ \text{m} < W_s <= 2\ \text{m}$ |
| MAV | 50 g < W <= 2 kg | $15\ \text{cm} < W_s <= 1\ \text{m}$ |
| NAV | 3 g < W <= 50 g | $2.5\ \text{cm} < W_s <= 15\text{cm}$ |
| CAV | 0.5 g < W <= 3 g | $0.25\ \text{cm} < W_s <= 2.5\text{cm}$ |
| SD | 0.005 g < W <= 0.5 g | $1\ \text{mm} < W_s <= 0.25\text{cm}$ |

Further, the design of UAV is classified into three categories such as fixed-wing, multirotor, and vertical take-off and landing (VTOL) (Macrina, Pugliese, Guerriero, & Laporte, 2020). Figure 3 presents the three types of design of UAVs. The above-mentioned three types of design of UAVs are briefly discussed as follows.

*Figure 3. Three categories of UAV based on its body types*

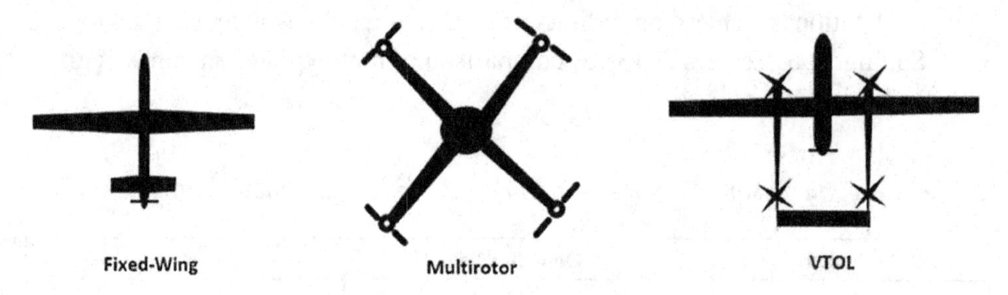


1. **Fixed wing:** UAVs with fixed wings have simpler structures, and they can carry greater payloads for long distances with less power consumption. However, this type of UAV is not fit for any kind of stationary operations like monitoring.
2. **Multirotor:** They can take off and land in a small place. They have high maintenance cost due to their structural complexity. Quadcopter, helicopter, hex copter, octocopter, etc. are some of these types of UAVs.
3. **Vertical take-off and landing (VTOL):** It is very flexible towards all types of works as it consists of the combined structure of both fixed-wing and multirotor types.

## Architecture of FANET

Based on connectivity the communication architecture of FANET is of two types: centralized and decentralized (Li, Zhou, & Lamont, 2013; Jiang & Han, 2018) . In a centralized architecture, the ground control station (GCS) acts as a gateway between any two UAVs. That means each UAV can directly communicate with GCS but needs GCS as an intermediate node for data transmission to other UAVs. On the other hand, decentralized architecture does not involve GCS in UAV to UAV data transmission. Figure 4 shows the centralized and decentralized architecture of FANET.

*Figure 4. (a) Centralized architecture and (b) Decentralized architecture of FANET*

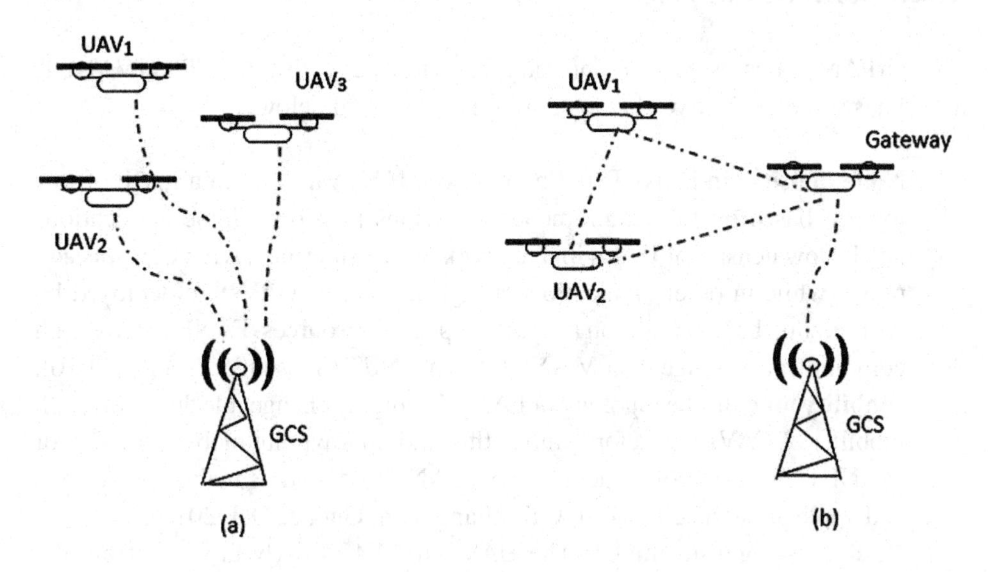


Decentralized architecture is of three types as mentioned in Table 4.

*Table 4. Decentralised architecture of UAV*

| Components of a Decentralized Network | UAV Ad-Hoc Network | Multi-Group UAV Ad-Hoc Network | Multi-Layer UAV Ad-Hoc Network |
|---|---|---|---|
| **Member UAVs** | **Number:** many member UAVs are present **Connectivity:** one member UAV to other member UAVs and member UAV to a gateway | **Number:** many groups of member UAVs are present **Connectivity:** one member UAV to other member UAVs of the same group and member UAV to a gateway | **Number:** many groups of member UAVs are present **Connectivity:** one member UAV to other member UAVs of the same group and member UAV to backbone UAV which connects a group member UAVs to a gateway |
| **Gateway** | **Number:** one UAV acts as the gateway **Connectivity:** member UAVs to gateway and gateway to GCS | **Number:** each group has a gateway **Connectivity:** a gateway is connected to its group member UAVs and GCS | **Number:** each group has a backbone UAV and from the backbone UAVs one is acted as gateway **Connectivity:** gateway is connected with all backbone UAVs and with GCS, where all backbone UAVs are also interconnected with each other |
| **GCS** | **Connectivity:** Gateway to GCS | **Connectivity:** GCS to every gateway | **Connectivity:** Gateway to GCS |

## Characteristics of FANET

Though FANET possesses many common characteristics with MANET and VANET, it also has some distinctive characteristics as mentioned below.

1. **Node density:** In FANET, the number of UAVs varies from a range of low to high based on the requirements of various missions. Some applications need a low density of UAVs in a network by considering a wide transmission range, while in other applications a high density of UAVs are deployed by minimizing the transmission range and specific resources. FANETs have high communication range than MANET and VANET (Yassein & Alhuda, 2016).
2. **Mobility model:** The topology of FANET is highly changeable due to the high mobility of UAVs. Therefore, unlike the random waypoint mobility model for MANET, the mobility model used in a FANET network is mainly predefined and application-specific (Lin, Cai, Zhang, Fan, Guo, & Dai, 2018).
3. **Radio propagation model:** The UAVs in a FANET fly far away from the ground, so the process of communication is vulnerable to the frequent changes in distance, blocking of line-of-sight by the obstacles, etc. Friis free space propagation model (Friis, 1946) is used in many simulation tests of FANET.
4. **Localization:** UAVs have a high speed of movement and in some other scenarios, they also move in an unpredictable manner. Thus, localization is an important factor to reach out to a UAV (Fadlullah, Takaishi, Nishiyama, Kato, & Miura, 2016).
5. **Resource constraint:** The main resource constraint of FANET is energy consumption. All UAVs are battery-powered, so they have a limited amount of energy resources. The consumption of energy mainly depends on the communication process, payloads, application-specific work, etc.

## Challenges of FANET

FANETs have many challenges and issues in various aspects (Srivastava & Prakash, Future FANET with application and enabling techniques: Anatomization and sustainability issues, 2021). Some of the issues and research direction based on that issues are mentioned in Table 5.

*Table 5. Different issues and various challenges of FANET*

| Issues | Challenges | Research Scope |
|---|---|---|
| Video data communication | Raised in routing overhead, a requirement of high bandwidth | Developing a routing mechanism to minimize the overhead, energy consumption and to maximize the throughput, etc. |
| Mobility model | Less performance on the result requirements of the applications and the data transmission | Designing a convenient mobility model for the improvement in result requirements and routing |
| Path planning | More requirement of time, energy, and incapable of proper data communication | Planning a suitable path on the basis of the target region, requirements of the application, and path length |
| Security issues | Malicious data transmission | Developing a routing protocol for secure data transmission |
| Faulty node | It leads to network failure, less network performance, faulty data transmission | Designing method for the identification and recovery of faulty UAVs. |

After the brief discussions about the background of VANET and FANET, the next section mentions various real-life applications of VANET.

## USAGE OF VANET IN REAL-LIFE APPLICATIONS

Although VANET faces several challenges, still it is widely used for various applications (Kumar, Mishra, & Chand, 2013). Various factors for the use of VANET for real-life applications are the availability of 5G for quick data transmission (Shahzad & Antoniou, 2019), availability of various communication standards like wireless in vehicular access environment (WAVE), dedicated short-range communications (DSRC) for effective communication among vehicles and RSU (Nampally & Sharma, 2018), successful use of MAC layer protocol i.e. IEEE 802.11p (Cao & Lee, 2020), presence of communication unit (OBU) in the vehicle (Arena & Pau, 2019), effective placement of RSU to maintain network connectivity (Huang, Li, & Zhang, 2018), etc. Many researchers have classified the applications of VANET into different categories. In (Ahmed, Pierre, & Quintero, 2017), the application of VANET is classified into two groups like safety and infotainment. VANET application is classified into traffic efficiency, road safety, and value-added applications in (Liang, Li, Zhang, Wang, & Bie, 2015). For the proposed chapter, the application of VANET is broadly classified into five categories. These are safety applications, infotainment applications, convenience applications, commercial applications, and productive applications. The next subsection discusses various categories of applications in brief.

## Safety Applications of VANET

The initial aim of VANET is to provide a safe transport application with the help of the intelligent transportation system (ITS) (Sheikh & Liang, 2019). To increase travel safety and to reduce accidents on roads, VANET transmits various information to the driver of the vehicle such as the presence of mobile and static obstacles on the street, broadcasting of road accident information for the diversion of the route, etc. Fire monitoring in urban regions is one of the safety applications in which researchers proposed for the quick transmission of fire location information to the nearest fire station and hospital. The above-mentioned application uses location service-based routing (Senapati, Khilar, & Swain, Fire controlling under uncertainty in urban region using smart vehicular ad hoc network, 2021). In order to avoid road traffic congestion, a traffic management application is proposed (Jayapal & Roy, 2016). The above application is used successfully through the global positioning system (GPS) and the smart mobile phone of all drivers is installed with a traffic management app for the quick communication of road traffic congestion information. Quick transmission of emergency data transmission is essential for the successful implementation of safety applications. For this author's proposed vehicle density-based emergency broadcast scheme (Tseng, Jan, Chen, Wang, & Li, 2010). To monitor patients' health before their arrival at the hospital, researchers proposed an approach called vehicular tele-medicines to exchange vehicular data from the ambulance to the hospitals (Mukhopadhyay & Raghunath, 2016).

## Infotainment Applications of VANET

Along with safety applications, VANET is also used for several infotainment applications. Researchers proposed multi-player mixed reality games for entertainment in the vehicular network (Sarakis, Orphanoudakis, Leligou, Voliotis, & Voulkidis, 2016). Infotainment applications demand the minimization of data transmission through flooding. For active navigation, researchers proposed one geometrical analysis to determine the position of the vehicle in order to avoid the broadcasting of the beacon signal (Salvo, Felice, Cuomo, & Baiocchi, 2012). Researchers have surveyed various ITS-based infotainment applications along with the multi constraint path problem (Oche, Tambuwal, Chemebe, Noor, & Distefano, 2020). For the implementation of the above infotainment applications, researchers have considered QoS-based routing protocols. Researchers have worked for the successful implementation of various comfort and entertainment applications like on-board Internet access, e-map downloading, etc. For the above-mentioned applications, researchers proposed the extension of IEEE 802.11p in order to support the multichannel operation of the WAVE architecture (Amadeo, Campolo, & Molinaro, 2012).

## Commercial Applications of VANET

Effective communication between seller and buyer by reducing the delay and waiting time is the objective of the commercial application of VANET. Due to its various commercial applications, VANET is popularly called a market on wheels. Researchers have proposed a location service-based routing protocol in order to minimize the communication network gap for the applications like buying and selling of products through vehicles (Bhoi, Puthal, Khilar, Rodrigues, Panda, & Yang, 2018). A virtual market called Flea Net has been proposed in which customers send the requirements through radio waves and sellers try to fulfil the request through the vehicular network (Lee, Lee, Park, & Gerla, 2009). Various advertisements such as the broadcast of shopping malls, offers of petrol pumps are locally propagated through VANET by focusing on the flexibility of the content sharing (Zhang, Zhao, & Cao, 2010). Wi-Fi sharing at the parking place among the vehicles is one of the commercial applications in which researchers proposed the enhancement of DSRC to support Wi-Fi sharing at the public parking place (Fitah, Badri, Moughit, & Sahe, 2018) .

## Convenience Applications of VANET

Providing comfort and convenience to the driver and passengers of the vehicle is the objective of the convenience application of VANET. Various convenience application through VANET helps to reduce waiting time, delay, and service time for the passenger of the vehicle. Automatic toll tax collection for the vehicles is proposed through VANET using routing protocol based on the parameters such as minimum density and shortest distance (Senapati, Khilar, & Sabat, An automated toll gate system using vanet, 2019). Avoidance of traffic congestion is another convenience application in which researchers have used the clustering approach for the avoidance of traffic congestion (Mohanty, Mahapatra, & Bhanja, 2019). Automatic parking service is one of the convenience applications for vehicle users. Using a vehicle to infrastructure communication, an available parking slot is communicated with the incoming vehicle to minimize the search time for the available slot (Senapati & Khilar, Automatic parking service through VANET: A convenience application., 2020). An automatic emergency vehicle like ambulance detection is one of the convenience applications to determine the location of the ambulance to reduce the search time to identify the location of the patients' house (Buchenscheit, Schaub, Kargl, & Weber, 2009).

### Productive Applications of VANET

To perform certain tasks easier and for the completion of tasks in less time, VANET provides a certain productive application. One of the productive applications is the monitoring of environmental parameters through VANET. Researchers proposed a probabilistic neural network to classify the environmental parameters into different zones (Senapati, Swain, & Khilar, Environmental monitoring under uncertainty using smart vehicular ad hoc network, 2020; Senapati, Khilar, & Swain, Environmental monitoring through Vehicular Ad Hoc Network: A productive application for smart cities, 2021). Secure automatic toll tax transaction is proposed through VANET (Chaurasia & Verma, 2014).

Thus, VANET is used for many real-life applications. Also, the availability of various meta-heuristic approaches also optimizes various performance metrics of VANET (Senapati & Khilar, Optimization of performance parameter for vehicular ad-hoc network (VANET) using swarm intelligence., 2020). Detection of faulty OBU automatically improves the overall performance of VANET (Senapati, Mohapatra, & Khilar, Fault Detection for VANET Using Vehicular Cloud., 2021; Senapati, Khilar, & Swain, Composite fault diagnosis methodology for urban vehicular ad hoc network., 2021).After the discussions of various real-life applications of VANET, the real-life applications of FANET are discussed in the next section.

## USAGE OF FANET IN REAL-LIFE APPLICATIONS

Recently FANETs are used in both military and civil domains. The autonomy, self-configurable, flexible, scalable, and cost-effectiveness features of the UAVs make a surge of attention towards the use of FANET in various ways. Some benefits of using a UAV system in real-life applications are: i) it can cover an extensive region efficiently, ii) work at any time like day or night for a long duration, iii) easily recoverable and comparatively economical, iv) capable of caring different payloads for different applications, and v) required less human involvement for the successful accomplishment of any missions (Swain, Khilar, & Senapati, 2022; Ambrosia & Zajkowski, 2015).Some key applications of FANET are search and rescue operation, forest fire detection, road traffic monitoring, crop management and monitoring, mail and delivery, etc. All the FANET based applications with detailed technologies and methods are discussed below.

## Search and Rescue Operation Using FANET

Alotaibi et. al has proposed a layered search and rescue process (LASR) based on the teamwork of a multi-UAV system. In a disaster area, many survivors are located in a particular region inside the total area, which is called the centre. The LSAR algorithm aims to focus more towards the centre, and gradually less towards the decreasing of distance from the centre. The total work process is divided into two phases, the first phase is the partitioning phase and the second is the actual search and rescue phase. A team of UAVs connected through a cloud server has partitioned the area into a set of layers, and the survivors located towards the centre have higher rescue priority than the survivors present in outer layers (Alotaibi, Alqefari, & Koubaa, 2019). A UAV-aided search and rescue process has been proposed for a GPS-disabled indoor environment. In this work, for location detection, a UAV uses the radio frequency signal generated from the smart devices of the victim. For a fast location tracking of the victim, it uses the reinforcement learning technique (Kulkarni, Chaphekar, Chowdhury, Erden, & Guvenc, 2020)]. Bejiga et. al has proposed an avalanche search and rescue operation using the aerial images of UAVs. The system model of SAR has consisted of the pre-processing and post-processing methods. In pre-processing, the features of the objects are extracted from the images using a convolutional neural network (CNN), and a trained support vector machine (SVM)is used at the top level of the CNN for the required object detection from the images. Next in post-processing, a classification is performed based on the Hidden Markov Model. Therefore, basically, the pre-processing stage focuses on the improvement of detection rate, however, the post-processing method focuses on the prediction performance of the classification (Bejiga, Zeggada, Nouffidj, & Melgani, 2016). Hayat et. al has proposed a method to plan an optimal path to cover the target region and to find the position of the victim in less operation time. Then it sends the location of the survivor to the ground control station for further communication establishment. A multi-objective function is designed for path length optimization, and the numerical method genetic algorithm is used for this optimization process (Hayat, Yanmaz, Brown, & Christian, 2017).

## Forest Fire Detection Using FANET

Belbachir et.al has proposed a method to detect the location of fire by a decision-based method. Usually, the target region for fire detection is an unknown environment for each UAV. Therefore, UAVs need to explore the entire region to detect the proper location of the forest fire. However, it is impractical for a UAV to cover the whole area because of its limited resources and vast forest region. In this paper, they have proposed a controlled navigation method for the movement of UAVs towards the location of a forest fire. The localization of forest fire is performed by a probabilistic method and it acted like an attractive force for the movement of UAVs. The working method has consisted of four modules such as forest-fire model, map updating model, exploration method, and finally vehicle model (Belbachir, Escareno, Rubio, & Sossa, 2015).Yuan et. al has proposed an image processing method for forest fire detection and tracking. This process has consisted of three steps such as fire search, fire confirmation, and fire observation. A sequence of image processing steps is executed for fire detection and tracking. In the first step, UAVs collect a set of images over a particular region, then a step of image pre-processing is performed for the noise reduction and the colour model conversion of all the collected images. In the next step, fire segmentation is performed using the process of thresholding. Morphological operation is executed for fire confirmation by eliminating the small irrelative objects. Finally, the blob counter mechanism is performed for object tracking (Yuan, Liu, & Zhang, 2015). Sudhakar et. al has designed a process of reducing the false alarm rate in detecting and monitoring forest fire. The proposed process includes three steps such as determination of colour code, identification of smoke motion, and fire categorization. A technique of forest fire detection by observing a set of images on the basis of both colour and movement inspection. It works using an agreeable state approximation technique, that evaluates re-iteratively the scenario of the alarms and the awareness by using the new estimations generated from other UAVs (Sudhakar, Vijayakumar, Kumar, Priya, Ravi, & Subramaniyaswamy, 2020). Jiao et. al has performed forest fire detection by using a deep learning technique. The high mobility, agility, and ability to fly in various altitudes make UAVs as suitable equipment for forest fire detection in a short period of time and at a lower cost. Conventional detection models are mainly used the RGB colour model. However, it provides a less precise result; therefore, this proposed method uses a YOLOv3 model based on UAV-areal images. The recognition algorithm uses a small-scale convolutional neural network (CNN) based on the YOLOv3 technique (Jiao, et al., 2019).

## Road Traffic Monitoring by FANET

Huang et. al has designed a method of road traffic monitoring based on a multi-UAV system. This process is accomplished by using four stages such as initial, searching, accumulating, and monitoring. Initially, a UAV starts searching the region until an accident occurs near that UAV. Then UAV moves to that particular position of accident for better observation and shares the accumulated information to its neighbours for further actions. This proposed method has the ability to monitor the road traffic as well as to keep an eye over the people in some events (Huang, Savkin, & Huang, Decentralized autonomous navigation of a UAV network for road traffic monitoring, 2021). A traffic monitoring system is proposed by using the 5G technology to detect the exceeding of the traffic speed limit and the violation of other traffic rules over a particular region, mostly the accident-prone region. In this method, UAVs first detect the vehicles that violate the speed and other safety parameters. Then it warns the driver of that vehicle for the first time, and at the second time, it shares all the information to the base station (Khan, Jhanjhi, Brohi, Usmani, & Nayyar, 2020). Byun et. al has proposed a method of traffic monitoring using deep learning technology. This method uses the captured images of UAVs to estimate the speed of the vehicles (Byun, Shin, Moon, Kang, & Choi, 2021). To provide real-time traffic congestion detection, an AI-based UAV-traffic monitoring system is proposed. This system considers the benefits of both the UAV system and AI for accurate recognition of traffic congestion. The monitoring part of this system uses UAVs to capture images over the assigned region, and the recognition system uses CNN for feature extraction and detection of congestion (Jian, Li, Yang, Wu, Ahmad, & Jeon, 2019).

## Crop Management and Monitoring Using FANET

Vega et. al has proposed a method of monitoring the sunflower crop based on the multi-temporal images of UAV. In the growing stage of sunflower temporal images are collected at different times of the day. The normalized difference vegetarian index (NDVI) is calculated based on the images for the estimation of the nitrogen, aerial biomass level (Vega, Ramirez, Saiz, & Rosua, 2015). Berni et. al has proposed a sensor-based approach for crop monitoring. In this process, a helicopter-based UAV uses the less cost thermal narrowband multispectral image sensors for capturing the images of crops (Berni, Zarco-Tejada, Su{\'a}rez, & Fereres, 2009). Wang et. al has developed a UAV-based visualization system for crop monitoring. Multi-rotor UAVs are used in this application. They have developed a visualization system based on the Django framework to collect and analyze the collected information about the crop field (Wang, Sun, Long, Zheng, Liu, & Li, 2018).

## Package Delivery Using UAV

Feng et. al has proposed a delivery method by UAVs using the markers in receiving location. In this method, UAV detects the colour of the marker by using its visual sensor analysis and lands at that particular location for the delivery of packages. They have used different colours of mailboxes as markers for the delivery purpose (Feng, Li, Ge, & Pan, 2020). Huang et. al has proposed a scheduling package delivery system by aerial drones. This system provides two types of delivery schemes. The first type is the drone-direct scheme, where a drone delivers the parcels directly to the customers, and the second type is the drone-vehicle collaborating scheme, where a drone delivers the parcels to customers by collaborating with other transport vehicles like the train, bus, etc. (Huang, Savkin, & Huang, Scheduling of a parcel delivery system consisting of an aerial drone interacting with public transportation vehicles, 2020). Grzybowski et. al has designed a UAV-based package delivery system, in which UAV uses the geographical information system based on QR address code to deliver the parcels (Grzybowski, Latos, & Czyba, 2020).

Looking at the popularity of VANET and FANET for various real-life applications, researchers have combined VANET and FANET for numerous applications. The next section discusses various real-life applications by the combination of VANET and FANET.

## USAGE OF THE COMBINATION OF VANET AND FANET FOR REAL-LIFE APPLICATIONS

VANET and FANET have some similarities in terms of the way of communication, smart vehicle, etc. and also they have differed in so many ways such as the mobility pattern, speed, flexibility, network topology, communication and sensing capabilities, etc. VANET and FANET are widely used separately in various real-life applications as discussed in the previous section. In different application domains, the utilization of VANET has certain advantages and challenges and so in the case of FANET. VANETs have specific limitations such as the restricted regions of mobility, technical issues in the vehicle deployment, connectivity challenges, etc. On the other hand, FANETs also have some restrictions such as energy constraints, high mobility, etc. that have not been the disadvantages in the case of VANET. Therefore, the combined use of both VANET and FANET complements each other's advantages and disadvantages for enhancing the overall performance of a mission. In recent days, researchers have found the idea of hybrid use as suitable for many real-life applications in terms of total execution time, effective data transmission, more area coverage, etc. The combining approach of VANET and FANET in some real-life applications is discussed as follows.

### Emergency Health Care Data Dissemination Through VANET and FANET

During the COVID-19 period, normally people are in a state of afraid to consult in the hospital. Researchers have combined the approach of FANET and VANET for the quick and emergency health data dissemination of the patient to the doctor present in the nearby hospital (Mukhopadhyay A. &., 2020). The approach is more suitable when the distance between the ambulance and hospital is more and the ambulance is not able to move due to traffic issues or due to natural calamity. The performance is evaluated in terms of parameters like delay, packet delivery ratio, throughput, etc.

## Improvement in the Performance of Clustering Routing Through VANET and FANET

Effective data routing is a challenging task for any ad hoc network. The high mobility of drones and vehicles creates a challenge among the researchers for an effective routing for FANET and VANET. Researchers have used various evolutionary algorithms based clustering routing like moth flame optimization, ant colony optimization, particle swarm optimization, Gray wolf optimization, etc, for increasing the performance of the network in terms of complex routing using both VANET and FANET (Tariq, 2020).

## Effective Reactive Routing for City Region Through VANET and FANET

Routing among the nodes for VANET in the city region for any applications is a challenging task. The main reason is the high mobility of the nodes and the presence of high buildings and trees creates the obstruction within the line of sight distance for effective communication. To overcome this problem, researchers have combined FANET and VANET to increase the performance of reactive routing for the city region (Sami Oubbati, 2020).

## Detection of Malicious Mobile Nodes in the Vehicular Network Using VANET and FANET

Effective communication among the nodes of VANET is essential for the successful implementation of any application. The presence of malicious nodes degrades the overall performance of VANET. Detection of malicious nodes in the VANET is essential. Researchers have used communication through drones and vehicles for the identification of malicious nodes in VANET (Fatemidokht, Rafsanjani, Gupta, & Hsu, 2021).

The overall summary of the technology used for various applications using VANET, FANET, and the combined approach of VANET and FANET is mentioned in Table 6.

*Table 6. Applications of VANET, FANET and the combination of VANET and FANET*

| Types of Ad Hoc Networks | Application Categories | Work Done |
|---|---|---|
| VANET | Safety applications | V2V communication among the mobile nodes of VANET to broadcast route diversion in the presence of any obstacles or any accident (Sheikh & Liang, 2019). |
| | | Automatic fire monitoring in the urban regions through location service-based routing (Senapati, Khilar, & Swain, Fire controlling under uncertainty in urban region using smart vehicular ad hoc network, 2021). |
| | | A traffic management system app is proposed to monitor the traffic using GPS and smartphone of vehicle users (Jayapal & Roy, 2016). |
| | | To monitor a patient's health before the arrival of the patient at the hospital is proposed using vehicular telemedicine approach (Mukhopadhyay & Raghunath, 2016). |
| | Infotainment applications | Sharing of resources in the vehicular network for multiplayer mixed reality games among the drivers of the vehicle in the nearby position (Sarakis, Orphanoudakis, Leligou, Voliotis, & Voulkidis, 2016). |
| | | Various ITS-based infotainment applications along with multi constraint path problems are considered with the focus on QoS-based routing protocols (Oche, Tambuwal, Chemebe, Noor, & Distefano, 2020). |
| | | Various comfort and entertainment applications like on-board Internet access, e-map downloading, etc. are successfully implemented (Amadeo, Campolo, & Molinaro, 2012). |
| | Commercial applications | Buying and selling of products through vehicles by minimizing the communication gap using the location service-based routing is proposed (Bhoi, Puthal, Khilar, Rodrigues, Panda, & Yang, 2018). |
| | | A virtual market called FleaNet has been proposed in which customers send the requirements through radio waves and sellers try to fulfil the request through VANET (Lee, Lee, Park, & Gerla, 2009). |
| | | Focusing on the flexibility of content sharing, advertisements are broadcasted for shopping malls, offers of petrol pumps, etc (Zhang, Zhao, & Cao, 2010). |
| | | Wi-Fi sharing at the public parking place is proposed by enhancing the features of the DSRC communication standard (Fitah, Badri, Moughit, & Sahe, 2018). |
| | Convenience applications | Automatic toll tax collection for the vehicles is proposed through VANET using routing protocol to minimize delay based on the density of vehicles and shortest distance (Senapati, Khilar, & Sabat, An automated toll gate system using vanet, 2019). |
| | | A clustering approach is used to avoid congestion of road traffic is proposed (Mohanty, Mahapatra, & Bhanja, 2019). |
| | | Using a vehicle to infrastructure communication, researchers have proposed an automatic parking service to minimize search time for available parking slots (Senapati & Khilar, Automatic parking service through VANET: A convenience application., 2020). |
| | Productive applications | Aautomatic environmental parameter monitoring is proposed by the researchers using a probabilistic neural network (Senapati, Khilar, & Swain, Environmental monitoring through Vehicular Ad Hoc Network: A productive application for smart cities, 2021). |
| | | Secure transaction during automatic toll tax collection is proposed (Chaurasia & Verma, 2014). |
| | | Performance improvement of VANET in terms of V2V communication is proposed by the automatic detection of faulty OBU (Senapati, Khilar, & Swain, Composite fault diagnosis methodology for urban vehicular ad hoc network., 2021). |

*Continued on following page*

*Table 6. Continued*

| Types of Ad Hoc Networks | Application Categories | Work Done |
|---|---|---|
| FANET | Search and rescue application | A layered search and rescue process (LASR) is performed by prioritizing the regions having more survivors than the other locations. So LASR is executed in two phases as partitioning and rescuing phases (Alotaibi, Alqefari, & Koubaa, 2019). |
| | | In a GPS-disabled environment, the radio frequency signals of smart devices and the reinforcement learning techniques are used for the location detection and tracking of a victim respectively (Kulkarni, Chaphekar, Chowdhury, Erden, & Guvenc, 2020). |
| | | An avalanche search and rescue operation uses CNN and SVM for the feature extraction and object detection from aerial images respectively, and the classification of objects is performed based on the Hidden Markov Model (Bejiga, Zeggada, Nouffidj, & Melgani, 2016). |
| | | To find the location of a victim, a path of optimized length is generated by using a multi-objective function, and the genetic algorithm is used for the process of optimization (Hayat, Yanmaz, Brown, & Christian, 2017). |
| | Forest fire detection and monitoring | Controlled navigation and a probabilistic method are used to reach out and localize the forest fire by a UAV respectively (Belbachir, Escareno, Rubio, & Sossa, 2015). |
| | | In an image processing method of fire detection and tracking following steps are executed. a) Image capturing, b) image pre-processing, c) fire segmentation, d) morphological operation for fire confirmation, and e) blob counter mechanism for fire tracking (Yuan, Liu, & Zhang, 2015). |
| | | Three steps as determination of colour code, identification of smoke motion, and fire categorization are used for reducing the false alarm rate in detecting and monitoring the forest fire (Sudhakar, Vijayakumar, Kumar, Priya, Ravi, & Subramaniyaswamy, 2020). |
| | | A deep learning recognition algorithm uses a small-scale convolutional neural network (CNN) based on the YOLOv3 technique to detect fire (Jiao, et al., 2019). |
| | Road traffic monitoring | A monitoring process of traffic is accomplished by using four stages such as initial, searching, accumulating, and monitoring (Huang, Savkin, & Huang, Decentralized autonomous navigation of a UAV network for road traffic monitoring, 2021). |
| | | UAV detects vehicles, that violate the road safety parameters by using the 5G technology (Khan, Jhanjhi, Brohi, Usmani, & Nayyar, 2020). |
| | | Deep learning is used in image processing for traffic monitoring (Byun, Shin, Moon, Kang, & Choi, 2021). |
| | | An artificial intelligence-based UAV system is also used for road traffic monitoring by feature extraction and object detection process (Jian, Li, Yang, Wu, Ahmad, & Jeon, 2019). |
| | Crop management and monitoring | The estimation of nitrogen and biomass level of sunflower crops is performed by using the multi-temporal images of UAVs (Vega, Ramirez, Saiz, & Rosua, 2015). |
| | | The less costly thermal narrowband multispectral image sensors are also used for capturing the images of the crop (Berni, Zarco-Tejada, Su{\'a}rez, & Fereres, 2009). |
| | | A visualization system based on the Django framework to collect and analyze the information about the crop field (Wang, Sun, Long, Zheng, Liu, & Li, 2018). |
| | Package delivery | The visual sensors of UAV detect the marker used for a delivery location for package delivery (Feng, Li, Ge, & Pan, 2020). |
| | | A drone-vehicle collaborating scheme delivers the parcels to customers by collaborating the drones with other vehicles like the train, bus, etc (Huang, Savkin, & Huang, Scheduling of a parcel delivery system consisting of an aerial drone interacting with public transportation vehicles, 2020). |
| | | UAV also uses the geographical information system based on QR address codes to deliver the parcels (Grzybowski, Latos, & Czyba, 2020). |

*Continued on following page*

*Table 6. Continued*

| Types of Ad Hoc Networks | Application Categories | Work Done |
|---|---|---|
| VANET & FANET combined approach | Emergency health data dissemination | Emergency health data dissemination for a person when the hospital is far from the ambulance and the ambulance is not able to move due to congestion or due to natural calamity is proposed through VANET and FANET (Mukhopadhyay A. &., 2020). |
| | Performance improvement of clustering routing | Using an evolutionary algorithm, the performance of cluster-based routing is enhanced by combining VANET and FANET (Tariq, 2020). |
| | Performance improvement of reactive routing. | The performance of the reactive routing is improved by avoiding the obstruction present in the line of sight distance by using the communication between the drones, vehicles, and RSU (Sami Oubbati, 2020). |
| | Detection of malicious nodes | Communication among the drones and vehicles is used for the detection of the malicious nodes of VANET (Fatemidokht, Rafsanjani, Gupta, & Hsu, 2021). |

## CONCLUSION AND FUTURE SCOPE

VANET and FANET act as the emerging research area for various real-life applications. In this chapter, the background of VANET is briefly discussed. In the background, the components, architecture, characteristics, and challenges of both VANET and FANET are mentioned. Numerous real-life applications using VANET, FANET, and the combined approach of VANET and FANET are briefly presented. In the future, the simulation and implementation of the VANET, FANET, and the combination of VANET and FANET could be performed for real-life applications.

## REFERENCES


Adão, T., Jonáš, H., Luís, P., José, B., Emanuel, P., & Raul, M. (2017). Hyperspectral imaging: A review on UAV-based sensors, data processing and applications for agriculture and forestry. *Remote Sensing*, *9*(11), 1110. doi:10.3390/rs9111110

Ahmed, H., Pierre, S., & Quintero, A. (2017). A flexible testbed architecture for VANET. *Vehicular Communications*, 115-126.

Al-Absi, M. A.-A., & Lee, H. (2021). Moving ad hoc networks—A comparative study. *Sustainability*, 61–87.

Alghamdi, Y., Munir, A., & La, H. M. (2021). Architecture, Classification, and Applications of Contemporary Unmanned Aerial Vehicles. *IEEE Consumer Electronics Magazine*, 9--20.

Alotaibi, E. T., Alqefari, S. S., & Koubaa, A. (2019). Lsar: Multi-uav collaboration for search and rescue missions. *IEEE Access: Practical Innovations, Open Solutions*, *7*, 55817–55832. doi:10.1109/ACCESS.2019.2912306

Amadeo, M., Campolo, C., & Molinaro, A. (2012). Enhancing IEEE 802.11 p/ WAVE to provide infotainment applications in VANETs. *Ad Hoc Networks*, *10*(2), 253–269. doi:10.1016/j.adhoc.2010.09.013

Ambrosia, V., & Zajkowski, T. (2015). Selection of appropriate class UAS/sensors to support fire monitoring: experiences in the United States. Handbook of Unmanned Aerial Vehicles, 2723-2754.

Arena, F., & Pau, G. (2019). *An overview of vehicular communications*. Future Internet. doi:10.3390/fi11020027

Bejiga, M. B., Zeggada, A., Nouffidj, A., & Melgani, F. (2016). A convolutional neural network approach for assisting avalanche search and rescue operations with UAV imagery. *Remote Sensing*, 100.

Bekmezci, I., Sahingoz, O. K., & Temel, Ş. (2013). Flying ad-hoc networks (FANETs): A survey. *Ad Hoc Networks*, *11*(3), 1254–1270. doi:10.1016/j.adhoc.2012.12.004

Belbachir, A., Escareno, J., Rubio, E., & Sossa, H. (2015). Preliminary results on UAV-based forest fire localization based on decisional navigation. In *Workshop on Research, Education and Development of Unmanned Aerial Systems (RED-UAS)* (pp. 377--382). IEEE. 10.1109/RED-UAS.2015.7441030

Berni, J. A., Zarco-Tejada, P. J., Suarez, L., & Fereres, E. (2009). Thermal and narrowband multispectral remote sensing for vegetation monitoring from an unmanned aerial vehicle. *IEEE Transactions on Geoscience and Remote Sensing*, *47*(3), 722–738. doi:10.1109/TGRS.2008.2010457

Bhoi, S. K., Puthal, D., Khilar, P. M., Rodrigues, J. J., Panda, S. K., & Yang, L. T. (2018). Adaptive routing protocol for urban vehicular networks to support sellers and buyers on wheels. *Computer Networks*, *142*, 168–178. doi:10.1016/j.comnet.2018.05.024

Buchenscheit, A., Schaub, F., Kargl, F., & Weber, M. (2009). A VANET-based emergency vehicle warning system. *IEEE Vehicular Networking Conference (VNC)* (p. 2009). IEEE.

Byun, S., Shin, I.-K., Moon, J., Kang, J., & Choi, S.-I. (2021). Road traffic monitoring from UAV images using deep learning networks. *Remote Sensing*, *13*(20), 4027. doi:10.3390/rs13204027

Cao, S., & Lee, V. C. (2020). An accurate and complete performance modeling of the IEEE 802.11 p MAC sublayer for VANET. *Computer Communications*, *149*, 107–120. doi:10.1016/j.comcom.2019.08.026

Chaurasia, B. K., & Verma, S. (2014). Secure pay while on move toll collection using VANET. *Computer Standards & Interfaces*, 403-411.

Chriki, A., Touati, H., Snoussi, H., & Kamoun, F. (2019). FANET: Communication, mobility models and security issues. *Computer Networks*, *163*, 106877. doi:10.1016/j.comnet.2019.106877

Fadlullah, Z. M., Takaishi, D., Nishiyama, H., Kato, N., & Miura, R. (2016). A dynamic trajectory control algorithm for improving the communication throughput and delay in UAV-aided networks. *IEEE Network*, *30*(1), 100–105. doi:10.1109/MNET.2016.7389838

Fatemidokht, H., Rafsanjani, M. K., Gupta, B. B., & Hsu, C.-H. (2021). Efficient and secure routing protocol based on artificial intelligence algorithms with UAV-assisted for vehicular ad hoc networks in intelligent transportation systems. *IEEE Transactions on Intelligent Transportation Systems*, *22*(7), 4757–4769. doi:10.1109/TITS.2020.3041746

Feng, K., Li, W., Ge, S., & Pan, F. (2020). Packages delivery based on marker detection for UAVs. In *Chinese Control and Decision Conference (CCDC)* (pp. 2094--2099). IEEE. 10.1109/CCDC49329.2020.9164677

Fitah, A., Badri, A., Moughit, M., & Sahe, A. (2018). Performance of DSRC and WIFI for Intelligent Transport Systems in VANET. *Procedia Computer Science*, *127*, 360–368. doi:10.1016/j.procs.2018.01.133

Friis, H. T. (1946). A note on a simple transmission formula. *Proceedings of the IRE*, 254-256. 10.1109/JRPROC.1946.234568

Goyal, A. K., Agarwal, G., & Tripathi, A. K. (2019). Network Architectures, Challenges, Security Attacks, Research Domains and Research Methodologies in VANET: A Survey. *International Journal of Computer Network & Information Security*, 37-44.

Grzybowski, J., Latos, K., & Czyba, R. (2020). *Low-cost autonomous UAV-based solutions to package delivery logistics*. Springer. doi:10.1007/978-3-030-50936-1_42

Hamdi, M. M., Audah, L., Rashid, S. A., Mohammed, A. H., Alani, S., & Mustafa, A. S. (2020). A review of applications, characteristics and challenges in vehicular ad hoc networks (VANETs). In *International Congress on Human-Computer Interaction, Optimization and Robotic Applications (HORA)* (pp. 1-7). IEEE.

Hassanalian, M., & Abdelkefi, A. (2017). Classifications, applications, and design challenges of drones: A review. *Progress in Aerospace Sciences*, *91*, 99–131. doi:10.1016/j.paerosci.2017.04.003

Hayat, S., Yanmaz, E., Brown, T. X., & Christian, C. (2017). Multi-objective UAV path planning for search and rescue. In IEEE international conference on robotics and automation (ICRA) (pp. 5569-5574). IEEE. doi:10.1109/ICRA.2017.7989656

Helen, D., & Arivazhagan, D. (2014). Applications, advantages and challenges of ad hoc networks. *Journal of Academia and Industrial Research*, 453–457.

Huang, H., Savkin, A. V., & Huang, C. (2020). Scheduling of a parcel delivery system consisting of an aerial drone interacting with public transportation vehicles. *Sensors (Basel)*, *20*(7), 20–45. doi:10.339020072045 PMID:32260583

Huang, H., Savkin, A. V., & Huang, C. (2021). Decentralized autonomous navigation of a UAV network for road traffic monitoring. *IEEE Transactions on Aerospace and Electronic Systems*, *57*(4), 2558–2564. doi:10.1109/TAES.2021.3053115

Huang, W., Li, P., & Zhang, T. (2018). RSUs placement based on vehicular social mobility in VANETs. In *IEEE Conference on Industrial Electronics and Applications (ICIEA)* (pp. 1255-1260). IEEE. 10.1109/ICIEA.2018.8397902

Jayapal, C., & Roy, S. S. (2016). Road traffic congestion management using VANET. In *International conference on advances in human machine interaction (HMI)* (pp. 1-7). IEEE.

Jian, L., Li, Z., Yang, X., Wu, W., Ahmad, A., & Jeon, G. (2019). Combining unmanned aerial vehicles with artificial-intelligence technology for traffic-congestion recognition: electronic eyes in the skies to spot clogged roads. *IEEE Consumer Electronics Magazine*, 81-86.

Jiang, J., & Han, G. (2018). Routing protocols for unmanned aerial vehicles. *IEEE Communications Magazine*, *56*(1), 58–63. doi:10.1109/MCOM.2017.1700326

Jiao, Z., Zhang, Y., Xin, J., Mu, L., Yi, Y., & Liu, H. (2019). A deep learning based forest fire detection approach using UAV and YOLOv3. In *1st International conference on industrial artificial intelligence (IAI)* (pp. 1--5). IEEE. 10.1109/ICIAI.2019.8850815

Khan, N. A., Jhanjhi, N., Brohi, S. N., Usmani, R. S., & Nayyar, A. (2020). Smart traffic monitoring system using unmanned aerial vehicles (UAVs). *Computer Communications*, *157*, 434–443. doi:10.1016/j.comcom.2020.04.049

Kulkarni, S., Chaphekar, V., Chowdhury, M. M., Erden, F., & Guvenc, I. (2020). UAV aided search and rescue operation using reinforcement learning. In *SoutheastCon* (pp. 1–8). IEEE. doi:10.1109/SoutheastCon44009.2020.9368285

Kumar, V., Mishra, S., & Chand, N. (2013). Applications of VANETs: present & future. *Communications and Network*, 12.

Lee, U., Lee, J., Park, J.-S., & Gerla, M. (2009). FleaNet: A virtual market place on vehicular networks. *IEEE Transactions on Vehicular Technology*, 344–355.

Li, J., Zhou, Y., & Lamont, L. (2013). Communication architectures and protocols for networking unmanned aerial vehicles. In IEEE Globecom Workshops (GC Wkshps) (pp. 1415-1420). IEEE.

Liang, W., Li, Z., Zhang, H., Wang, S., & Bie, R. (2015). Vehicular ad hoc networks: Architectures, research issues, methodologies, challenges, and trends. *International Journal of Distributed Sensor Networks*, *11*(8), 745303. doi:10.1155/2015/745303

Lin, J., Cai, W., Zhang, S., Fan, X., Guo, S., & Dai, J. (2018). A Survey of Flying Ad-Hoc Networks: Characteristics and Challenges. In *Eighth International Conference on Instrumentation & Measurement, Computer, Communication and Control (IMCCC)* (pp. 766--771). IEEE. 10.1109/IMCCC.2018.00165

Macrina, G., Pugliese, L. D., Guerriero, F., & Laporte, G. (2020). Drone-aided routing: A literature review. *Transportation Research Part C, Emerging Technologies*, *120*, 102762. doi:10.1016/j.trc.2020.102762

Mohanty, A., Mahapatra, S., & Bhanja, U. (2019). Traffic congestion detection in a city using clustering techniques in VANETs. *Indonesian Journal of Electrical Engineering and Computer Science*, 884-891.

Mukhopadhyay, A. (2020). FANET based Emergency Healthcare Data Dissemination. In *Second International Conference on Inventive Research in Computing Applications (ICIRCA)* (pp. 170-175). IEEE.

Mukhopadhyay, A., & Raghunath, S. (2016). Feasibility and performance evaluation of VANET techniques to enhance real-time emergency healthcare services. In *International Conference on Advances in Computing, Communications and Informatics (ICACCI)* (pp. 2597--2603). IEEE. 10.1109/ICACCI.2016.7732449

Nampally, V., & Sharma, M. R. (2018). Information sharing standards in communication for VANET. *International Journal of Scientific Research in Computer Science Applications and Management Studies*, 2319–1953.

Oche, M., Tambuwal, A. B., Chemebe, C., Noor, R. M., & Distefano, S. (2020). VANETs QoS-based routing protocols based on multi-constrained ability to support ITS infotainment services. *Wireless Networks*, *26*(3), 1685–1715. doi:10.100711276-018-1860-7

Salvo, P., Felice, M. D., Cuomo, F., & Baiocchi, A. (2012). Infotainment traffic flow dissemination in an urban VANET. In *IEEE Global Communications Conference (GLOBECOM)* (pp. 67-72). IEEE. 10.1109/GLOCOM.2012.6503092

Sami Oubbati, O. C., Chaib, N., Lakas, A., Bitam, S., & Lorenz, P. (2020). U2RV: UAV-assisted reactive routing protocol for VANETs. *International Journal of Communication Systems*, *33*(10), 4104. doi:10.1002/dac.4104

Sarakis, L., Orphanoudakis, T., Leligou, H. C., Voliotis, S., & Voulkidis, A. (2016). Providing entertainment app lications in VANET environments. *IEEE Wireless Communications*, *23*(1), 30–37. doi:10.1109/MWC.2016.7422403

Senapati, B. R. (2021). Composite fault diagnosis methodology for urban vehicular ad hoc network. *Vehicular Communications*, 100337.

Senapati, B. R., & Khilar, P. M. (2020). Automatic parking service through VANET: A convenience application. In Progress in Computing, Analytics and Networking (pp. 151-159). Springer. doi:10.1007/978-981-15-2414-1_16

Senapati, B. R., & Khilar, P. M. (2020). Optimization of performance parameter for vehicular ad-hoc network (VANET) using swarm intelligence. In Nature Inspired Computing for Data Science (pp. 83-107). Springer.

Senapati, B. R., Khilar, P. M., & Sabat, N. K. (2019). An automated toll gate system using vanet. In *IEEE 1st international conference on energy, systems and information processing (ICESIP)* (pp. 1-5). IEEE.

Senapati, B. R., Khilar, P. M., & Swain, R. R. (2021). Composite fault diagnosis methodology for urban vehicular ad hoc network. *Vehicular Communications*, 100337.

Senapati, B. R., Khilar, P. M., & Swain, R. R. (2021). Environmental monitoring through Vehicular Ad Hoc Network: A productive application for smart cities. *International Journal of Communication Systems*, 4988.

Senapati, B. R., Khilar, P. M., & Swain, R. R. (2021). Fire controlling under uncertainty in urban region using smart vehicular ad hoc network. *Wireless Personal Communications*, *116*(3), 2049–2069. doi:10.100711277-020-07779-0

Senapati, B. R., Mohapatra, S., & Khilar, P. M. (2021). Fault Detection for VANET Using Vehicular Cloud. In *Intelligent and Cloud Computing* (pp. 87–95). Springer. doi:10.1007/978-981-15-6202-0_10

Senapati, B. R., Swain, R. R., & Khilar, P. M. (2020). Environmental monitoring under uncertainty using smart vehicular ad hoc network. In *Smart intelligent computing and applications* (pp. 229–238). Springer. doi:10.1007/978-981-13-9282-5_21

Shahzad, M., & Antoniou, J. (2019). Quality of user experience in 5G-VANET. In *IEEE 24th international workshop on computer aided modeling and design of communication links and networks (camad)* (pp. 1-6). IEEE.

Sheikh, M. S., & Liang, J. (2019). A comprehensive survey on VANET security services in traffic management system. *Wireless Communications and Mobile Computing*, *2019*, 1–23. doi:10.1155/2019/2423915

Singh, S. K. (2015). A comprehensive survey on fanet: Challenges and advancements. *International Journal of Computer Science and Information Technologies*, 2010–2013.

Srivastava, A., & Prakash, J. (2021). Future FANET with application and enabling techniques: Anatomization and sustainability issues. *Computer Science Review*, *39*, 100359. doi:10.1016/j.cosrev.2020.100359

Sudhakar, S., Vijayakumar, V., Kumar, C. S., Priya, V., Ravi, L., & Subramaniyaswamy, V. (2020). Unmanned Aerial Vehicle (UAV) based Forest Fire Detection and monitoring for reducing false alarms in forest-fires. *Computer Communications*, *149*, 1–16. doi:10.1016/j.comcom.2019.10.007

Sun, W. L., Liu, J., & Zhang, H. (2017). When smart wearables meet intelligent vehicles: Challenges and future directions. *IEEE Wireless Communications*, *24*(3), 58–65. doi:10.1109/MWC.2017.1600423

Swain, S., Khilar, P. M., & Senapati, B. R. (2022). An effective data routing for dynamic area coverage using multidrone network. *Transactions on Emerging Telecommunications Technologies*, *33*(9), 4532. doi:10.1002/ett.4532

Tariq, R. I., Iqbal, Z., & Aadil, F. (2020). IMOC: Optimization technique for drone-assisted VANET (DAV) based on moth flame optimization. *Wireless Communications and Mobile Computing*, *2020*, 860646. doi:10.1155/2020/8860646

Tseng, Y.-T., Jan, R.-H., Chen, C., Wang, C.-F., & Li, H.-H. (2010). A vehicle-density-based forwarding scheme for emergency message broadcasts in VANETs. In *The 7th IEEE International Conference on Mobile Ad-hoc and Sensor Systems (IEEE MASS 2010)* (pp. 703-708). IEEE.

Ur Rehman, S., Khan, M. A., Zia, T. A., & Zheng, L. (2013). Vehicular ad-hoc networks (VANETs)-an overview and challenges. *Journal of Wireless Networking and Communications*, 29-38.

Vega, F. A., Ramirez, F. C., Saiz, M. P., & Rosua, F. O. (2015). Multi-temporal imaging using an unmanned aerial vehicle for monitoring a sunflower crop. *Biosystems Engineering*, *132*, 19–27. doi:10.1016/j.biosystemseng.2015.01.008

Wang, X., Sun, H., Long, Y., Zheng, L., Liu, H., & Li, M. (2018). Development of visualization system for agricultural UAV crop growth information collection. *IFAC-PapersOnLine*, *51*(17), 631–636. doi:10.1016/j.ifacol.2018.08.126

Yassein, M. B., & Alhuda, N. (2016). Flying ad-hoc networks: Routing protocols, mobility models, issues. *International Journal of Advanced Computer Science and Applications*.

Yuan, C., Liu, Z., & Zhang, Y. (2015). UAV-based forest fire detection and tracking using image processing techniques. In *International Conference on Unmanned Aircraft Systems (ICUAS)* (pp. 639--643). IEEE. 10.1109/ICUAS.2015.7152345

Zhang, J., Chen, T., Zhong, S., Wang, J., Zhang, W., & Zuo, X. (2019). Aeronautical Ad Hoc Networking for the Internet-Above-the-Clouds. *Proceedings of the IEEE*, 868-911. 10.1109/JPROC.2019.2909694

Zhang, Y., Zhao, J., & Cao, G. (2010). Roadcast: A popularity aware content sharing scheme in vanets. *Mobile Computing and Communications Review*, *13*(4), 1–14. doi:10.1145/1740437.1740439

# Chapter 4
# Establishment of FANETs Using IoT-Based UAV and Its Issues Related to Mobility and Authentication

**Uma Mageswari R.**
*Vardhaman College of Engineering, India*

**Murugan K.**
*Bannari Amman Institute of Technology, India*

**Nallarasu Krishnan**
*Tagore Engineering College, India*

**Sankar Ram C.**
*Anna University, India*

**Mohammed Sirajudeen Yoosuf**
*VIT-AP University, India*

## ABSTRACT

*The tremendous evolution of wireless communication as well as the drastic adoption of technology by the latest computing devices known to be IoT, makes it possible for emerging applications to providing ubiquitous services. This technique transformed the quality of present lifestyle of the people. When compared with all other technologies, the mobile adhoc networks become widely adapted in many fields because of the non-requirement of centralized infrastructure support. Adopting this nature, it became easy to establish networks like WSN and also form networks using IoT devices. As FANET (flying/fast adhoc network) is known for its mobility and instant formation of network with the help of available nodes within its communication range, there is a great challenge related to mobility and authenticity of the participating devices by exempting malicious nodes. FANETs incorporate unmanned aerial vehicles and drones as a part of their communication networks. In this chapter, deployment of IoT-based FANETs along with mobility and security is handled.*

DOI: 10.4018/978-1-6684-3610-3.ch004

## INTRODUCTION

Basically, wireless networks are categorised into infrastructure-based and infrastructure-less networks. To avoid the limitations and challenges faced by infrastructure-based networks related to range and other support constraints, the concept of UAV came to exist which works as same as ad hoc fashion and support to form a network known to be FANET (Lakew et al., 2020). Flying Ad-Hoc Networks (FANETs) are the networks formed with the nodes that support high mobility in air and the network formation is adopted from the existing ad hoc networking structure. The devices participating in the network formations are usually known to be unmanned aerial vehicles (UAVs) in the form of drones which are known to be a form of Internet of Things (IoT) device. Due to the advancement in the field of electronic, sensor and communication technologies, the design of IoT based UAV systems are playing their roles in various fields. These IoT based UAV can fly unconventionally as well as function remotely without need for any human intervention. Because of features like versatility, flexibility, easy installation and relatively small operating expenses, the usage of UAVs promises new ways for both military and civilian applications, such as search and destroy operations, border surveillance, disaster monitoring, remote sensing, and traffic monitoring etc. Nowadays multi-UAV are used for performing tasks like monitoring and have the capability to work in the corporate fashion (Bekmezci et al., 2013). The figure 1 shows the transformation of MANET to VANET and to the existing FANET.

*Figure 1. Adaptation of MANET towards VANET to FANET*

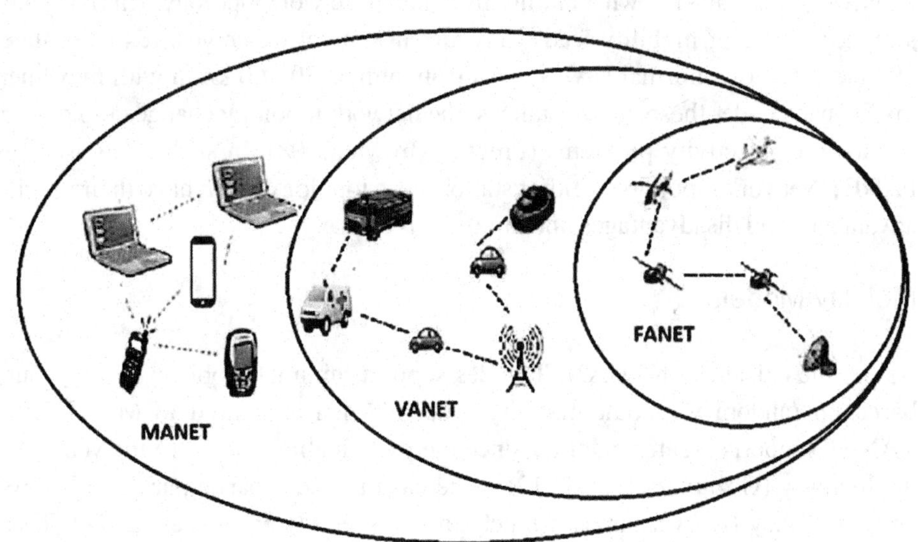

The advantages of the multi-UAV systems can be summarized as follows:

- Cost: The acquisition and maintenance cost of small UAVs is much lower than the cost.
- Scalability: Easy to add up the UAV thereby enables increased coverage ease of operation.
- Survivability: Even if any one of the participating UAV fails to operate the other available UAV does the operation of the down UAV.
- Speed-up: The focused missions can be completed faster with a higher number of UAVs.
- Small radar cross-section: Instead of one large radar cross-section, multi-UAV systems produce very small radar cross-sections, which is crucial for military applications

Despite these idiosyncratic traits, they brand the FANETs as a suitable resolution for diverse circumstances, but still the issues related to communications and networking of the multiple UAVs need to be focused (Bekmezci et al., 2013).

## FANETs and Its Characteristics

The following are the characteristics support by the FANETs

### Network Topology

FANETs in general is known for its highly dynamic network topology. This is because a higher degree of mobility leads to recurrent topology changes. It's been stated that the speed of a normal UAV ranges from almost 30-460 km/h with movement to 3D-space under these circumstances, the network topology changes rapidly that results in connectivity problems (Frew & Brown, 2008). In such circumstances, FANET Network topology is either star or mesh topology. Both have their specific advantages and disadvantages and are used as such.

### Mobility Models

As per the existence the MANET nodes support minimal mobility in the ground based on random waypoint mobility model. When compared to MANET, the VANET supports greater mobility since the participating nodes are the vehicle on the highway (Gozalvez et al., 2012). Whereas the nodes participating in FANET are UAVs they fly over air with much greater speed (Chriki et al., 2019). These FANET with multi-UAV supported applications, following global path plans are generally preferred, where the movements are based on a predetermined path and reflects a regular mobility model. In autonomous multi-UAV systems, the flight plan is not predetermined. Semi-Random Circular Movement (SRCM) mobility model is proposed for the fast and sharp UAV movements under predefined flight plans. In a random UAV movement model, each UAV decides on its movement direction according to a predefined Markov process. The following figure shows some of the mobility models supported by FANET.

*Figure 2. FANET mobility models*

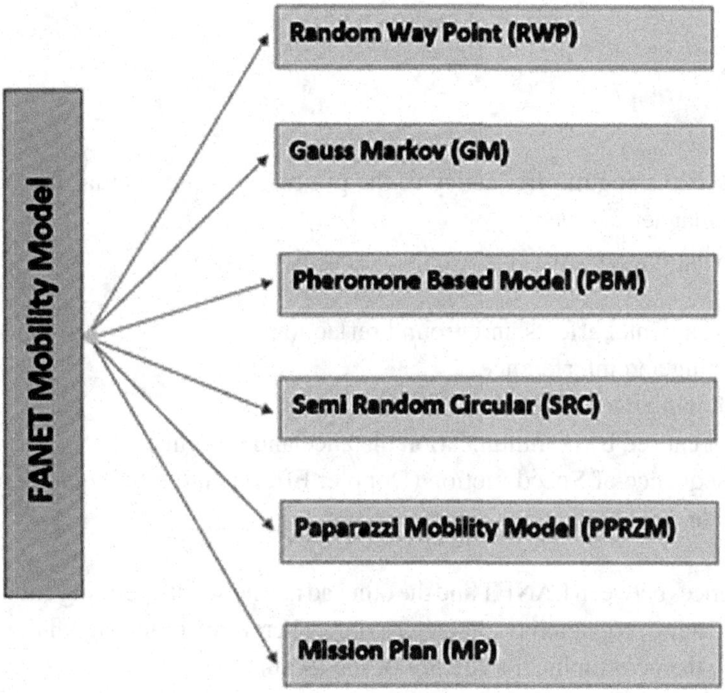

## Node Density

Node density is described as the distribution of the number of nodes in a unit area. FANET are created by the nodes in the sky, each node usually UAVs participating in this network are mostly several kilometres apart from each other. Hence it is stated that FANET node density is much lower when related with the MANET and VANET. Also, these node densities will vary based upon the scenario of usage and requirements.

## Radio Propagation Model

Radio waves radiate from the transmitter antenna and then propagate through environments then reach the receiver antenna (Jasim et al., 2020). During propagation the radio wave gets reflected, scattered, and diffracted by terrain, buildings, and other objects. In fact, these characteristics depend upon the distance between the two antennas, the path(s) taken by the signal, and the environment (buildings and other objects) around the path. The radio wave propagation and its characteristics can be expressed as mathematical functions that are called radio propagation modelling (Ding, 2019).

$$n = 1 - \frac{{}^{2}k}{2} \quad (1)$$

Where, ωk is the angular frequency of the plasma; ω is the angular frequencies of the electromagnetic wave
The following features affect the radio propagation:

1. Environmental effects and ground reflection
2. Jamming and interference
3. Shadowing due to the equipment in UAV
4. Effect caused by communication distance and mobility
5. Consequence of Speed motion (Doppler Effect) on the link quality and UAV attitude

Differences between FANET and the other ad hoc network operating environments affect the radio propagation characteristics. Therefore, radio signals are mostly affected by the geographical structure of the terrain.

## Localization

The main aim behind the localization is to identify the exact location of the nodes participating in the communication known to be geospatial localization. At present for the localization, the available methods are global positioning system (GPS), beacon (or anchor) nodes, and proximity-based localization. The known fact is that UAV operates on high mobility therefore needs a localization technique that predicts the location in smaller time intervals. The duration taken by the GPS in general is one-second interval, and for certain FANET protocols needs less than this interval. In order to support this, each UAV must be equipped with a GPS and an inertial measurement unit (IMU). This helps in contributing the position to the other UAVs at any time. Thus, by the calibration of the IMU by the GPS signal, the position of the UAV can be obtained at an earlier rate (Kheli et al., 2018).

### Energy Management

The major issue that incumbents the potency of a UAV system is the restricted on-board energy support (Khan et al., 2019). Network lifetime is a key design subject for FANET which can be improved by incorporating energy efficient communication protocols. It's also suggested that network lifetime can be enhanced by energy-efficient deployment and energy-efficient operations. Energy efficiency is addressed by a scheduling framework for cooperative UAVs communication (Tran et al., 2017).

## FANETs Mobility Model

The mobility models are categorized into five classes (Bujari et al., 2017):

1. Randomized mobility models
2. Time/space dependent mobility models
3. Path-planned mobility models
4. Group mobility models
5. Topology-control based mobility models.

### Randomized Mobility Models

Randomized mobility represents multiple mobile nodes whose actions are completely independent of each other and past actions. This model is further classified as Random Walk (RW), Random Waypoint (RWP) and Random Direction (RD), these models support the nodes to randomly choose a direction (or a waypoint) and speed to move for a certain period. Another form of randomized mobility model is Manhattan Grid (MG), that adopts a grid road topology where the node is allowed to move in horizontal or vertical direction on an urban map.

### Time/Space Dependent Mobility Models

This mobility model tries to avoid sharp speed and sharp direction changes. This model is further classified as Boundless Simulation Area (BSA) and Gauss-Markov (GM), this makes use of the association between the previous and the current values related to direction as well as speed and generates the new values at every instance. The other classification under this model is the Smooth Turn (ST) mobility model, which allows the mobile nodes to move in curved trajectories. It allows to choose a point in space, based on that point UAV will be circling around it, until another turning point is selected.

### Path-Planned Mobility Models

These mobility models offer a path scheme which uses a predefined shape, the UAVs follow this scheme and randomly change to another pattern or repeat the same one. This model is further classified as Semi-Random Circular Movement (SRCM), which is designed for the curved movement scenarios of UAVs. The Paparazzi mobility model (PPRZM) is a stochastic mobility model that is based on a state machine containing six movement pattern states, each UAV chooses any one of these patterns and at a random speed.

### Group Mobility Models

This model holds a spatial constraint among all the mobile nodes. This model is further classified as Reference Point Group mobility (RPGM) which deals with the random motion of mobile nodes around a reference point that moves on the area with a simple RWP model.

### Topology-Control Based Mobility Models

The mission constraints application needs the mobility model that has the control of mobile nodes topology. This model is further classified as Distributed Pheromone Repel (DPR) mobility model which supports investigation and search missions, here each mobile node maintains an own pheromone map. The Self-Deploy Point Coverage (SDPC) mobility model is used for disaster scenarios, where a set of UAVs are deployed in order to create a communication infrastructure in areas the required monitoring (Lin et al., 2019).

## Layers in FANETs

The FANETs uses the layer structure based on traditional ISO/OSI or TCP/IP model which holds various layers like Physical layer, Link layer, Network layer, Transport layer and Application layer

## Physical Layer

As per ISO/OSI model Physical layer deals with the hardware implementation. It specifies the signal transmission technologies such as encoding and moves of the data bits over the physical medium using different types of antennas and the characteristics of links as well as channels used to establish the connectivity with the neighbors (Sawalmeh & Othman, 2021).

Radio propagation models and antenna structures both are considered as (Vasilyev et al., 2020) the key factors affecting the physical level of FANET are the multipath model of radio waves and the structure of antennas (Bekmezci et al., 2013). The physical layer design supported by UAV uses the following key factors.

1. Type of modulation
2. Data transfer rate
3. Quality of the radio signal
4. Energy efficiency
5. Weight and size indicators.

## Mac Layer

The link layer is also known to be the Mac (Media Access Control) layer that takes care of the channel handling and allocation based on IEEE 802.15.4. As per the functionality channels are considered as a streaming channel for data support and the control channel for sending the control information, which have significantly different requirements and must be considered separately (Vasilyev et al., 2020).

Adaptive MAC protocol provides the solution for link quality fluctuations caused by the high mobility of nodes participating in FANET. This causes endlessly varying the distance between the UAV nodes and also introduces latency. The Adaptive MAC protocol uses an omnidirectional antenna for control packet transfer and directional antenna for data packets transfer which proves that End to End Delay, Throughput and Bit Error Rate were improved (Alshbatat & Dong, 2010). A Token based approach known as Token MAC provides a solution to overcome the problem in traditional contention-based protocols and link failures due to high mobility by updating target information (Cai et al., 2012).

LODMAC (Location Oriented Directional MAC) is a novel MAC protocol that integrates the directional antennas and location estimation of the neighboring nodes within the MAC layer. The well-known directional deafness problem is handled by incorporating the Busy to Send (BTS) packet along with the existing Request to Send (RTS) and Clear to Send (CTS) packets, this addresses the LODMAC as an efficient Protocol.

Figure 3. Phases in LODMAC

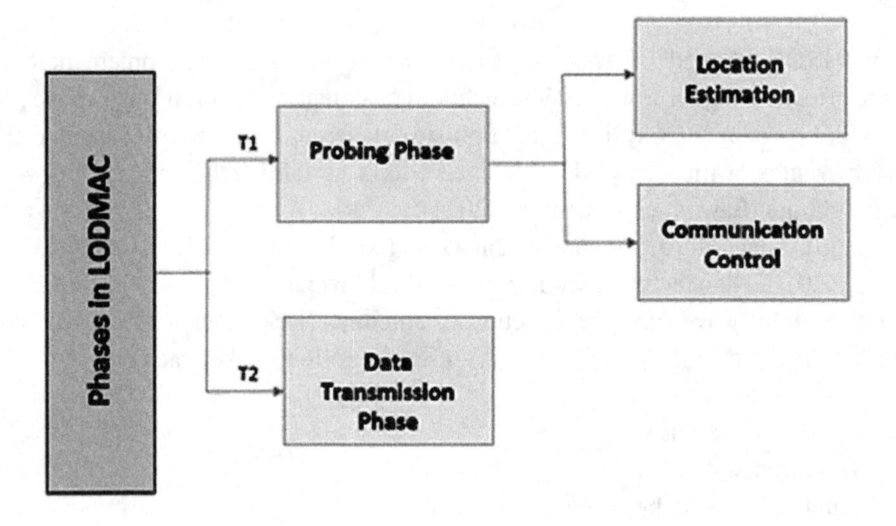


The LODMAC protocol utilizes two parallel phases namely, the probing phase and the data transmission phase respectively. The probing phase is divided into two slots as shown in the above figure as the location estimation slot and the communication control slot. In LODMAC, one of the transceivers, T1 is assigned for the probing phase and the T2 is assigned for the data transmission phase. The probing phase lasts for 1s exactly which is equal to the global GPS interval and in this phase, UAV nodes disseminate their location information to the neighboring nodes and exchange communication control packets. Separating the control and data channel results in increased network capacity as well as suppresses the interference among the two antennas. These antennas work in different frequencies and are physically installed on the UAVs. The LODMAC protocol achieves high throughput, maximizes utilization, minimizes network delay when compared with the existing well-known DMAC (Directional MAC) protocol designed for FANET MAC protocols (Temel & Bekmezci, 2015).

## Network Layer

The main functionality of network layer is related with unique addressing and routing. Finding the optimal path is a challenging task in the FANET environment since the dynamic change in the network topology as the individual noes get disconnected due to its movement. The following figure shows the general routing protocols available in network layer used by Adhoc network which can be adopted to the FANET.

The classification of FANET network layer protocol is shown in Figure 4.

*Figure 4. Classification of Routing Protocol*

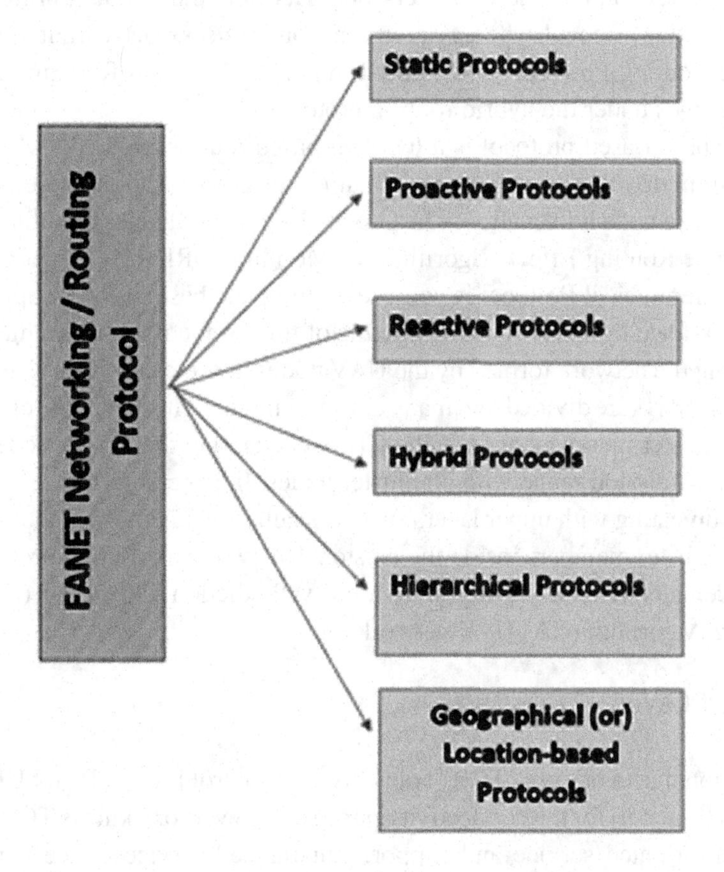


In static protocol the routing table is fixed throughout the operation, the changes in the node i.e. UAV position cannot be reflected in the table. It holds the various routing protocols like Load Carry and Deliver Routing (LCAD), Data Centric Routing (DCR), Multi-Level Hierarchical Routing (MLH). As the FANET dynamically changes its topology this static routing doesn't suit. The proactive protocol maintains the routing table that holds all the latest updates about the neighbouring UAV nodes. The Optimized Link State Routing (OLSR), Directional Optimized Link State Routing, Destination Sequenced Distance Vector comes under the Proactive protocol. The drawback in this routing is bandwidth not being used efficiently since a lot of data will be exchanged between the nodes for updating the table. The Reactive routing protocol is an on-demand routing protocol which is a bandwidth efficient protocol. This protocol has Dynamic Source Routing (DSR), Ad-hoc On-demand Distance Vector and Time-slotted On-demand Routing. Hybrid routing protocol is a combination of reactive and proactive protocols that is suitable to be used in large networks. The participating network is divided into a number of zones, namely, intra-zone and inter-zone. Here the intra-zone routing uses the proactive routing protocol, whereas inter-zone routing uses reactive routing protocol. The Zone Routing Protocol (ZRP) and Temporarily Ordered Routing Algorithm (TORA) comes under the hybrid routing protocol.

Geographic Based protocol is a location-based routing protocol, which works by the assumption that sender knows about the receiver location, there is no need to maintain the node information. The Greedy Perimeter Stateless Routing (GPSR) and Distance Routing Effect Algorithm for Mobility (DREAM) comes under this protocol. Hierarchical Protocol is considered to be the best protocol supported for the network form by UAV. As per the name of this protocol it forms a hierarchical level among the network formed by the UAV nodes, it serves for large size FANET. Thus the FANETs are divided into many clusters and each cluster holds a cluster head (CH) and cluster members nodes. The CH is one among the cluster nodes, which is in the transmission range with all cluster nodes. In general CH takes care about the communicating with upper layers such as satellites and UAVs and it broadcasts data to its cluster members inside the cluster. The two main routing protocols that come under hierarchical routing are Mobility Prediction Clustering (MPC) and Clustering Algorithm (CA) UAV networking.

## Transport Layer

The traditional network uses TCP (Transmission Control Protocol) and UDP (User Datagram Protocol) for the connectivity purpose. As everyone knows TCP provides connection-oriented service and supports guaranteed services since it promotes retransmission whereas UDP provides connectionless service and is unreliable. But due to fluid topology, disconnections and rapid changes in links between UAVs in the FANETs, traditional transport layer protocols such as User Datagram Protocol (UDP) and Transmission Control Protocol (TCP) don't directly suit FANETs.

To address the performance issues of traditional TCP such as long delay and reliable connection for packet transmissions and to suit the space environment, the enhanced protocols were proposed (Wang et al., 2009). Space Communication Protocol Standard - Transport Protocol (SCPS-TP), Satellite Transport Protocol (STP), eXtended STP (XSTP), TCP Peach, TP-Planet, and TCP Westwood (TCPW) are some the enhance TCP that support UAV network to do communication flawlessly.

## Application Layer

The Application layer specifies how to interface with the bottom layer protocol for data transmission. The common application layer protocols which support FANETs are HyperText Transfer Protocol (HTTP), Constrained Application Protocol (CoAP),

Message Queuing Telemetry Transport (MQTT), Extensible Messaging and Presence Protocol (XMPP), Websocket, Data Distribution Service (DDS), Advanced Message Queuing Protocol (AMQP).

HTTP laid the base for World Wide Web, which is a stateless protocol that supports commands like GET, PUT, POST, DELETE etc. It works under the communication model Request response that operates using port 80. CoAP is a M2M (Machine to Machine) application support protocol runs on the top of the UDP and supports constrained environment with restricted devices and network. CoAP design interface with the HTTP protocol in easy manner. Websocket uses TCP for connectivity thus allows the flow of streams of messages over the established socket, where the client can be any form of devices including IoT devices.

MQTT known to be light weight messaging protocol which adapts publish-subscribe model can handle unreliable network with minimal resource requirements. XMPP is an application layer protocol that supports real time communication and streaming much supportive for IoT devices.

## Routing Technique

Routing is defined as choosing the path between the end points using the intermediate nodes in the network. The best route is predicted by the metrics like packet drop rate, throughput, packet receiver ratio, end to end delay etc. Due to the nature of high mobility of FANET the path established breaks frequently, thus it is challenging to establish path frequently. Routing protocols are grouped under topology-based routing, hierarchical routing, Swarm intelligence base routing, heterogeneous based routing, position-based routing (Khan et al., 2020). The following figure shows the routing protocol taxonomy.

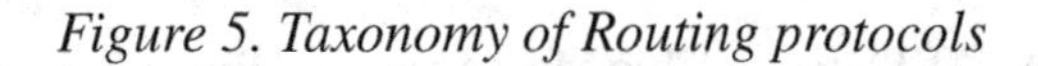

*Figure 5. Taxonomy of Routing protocols*

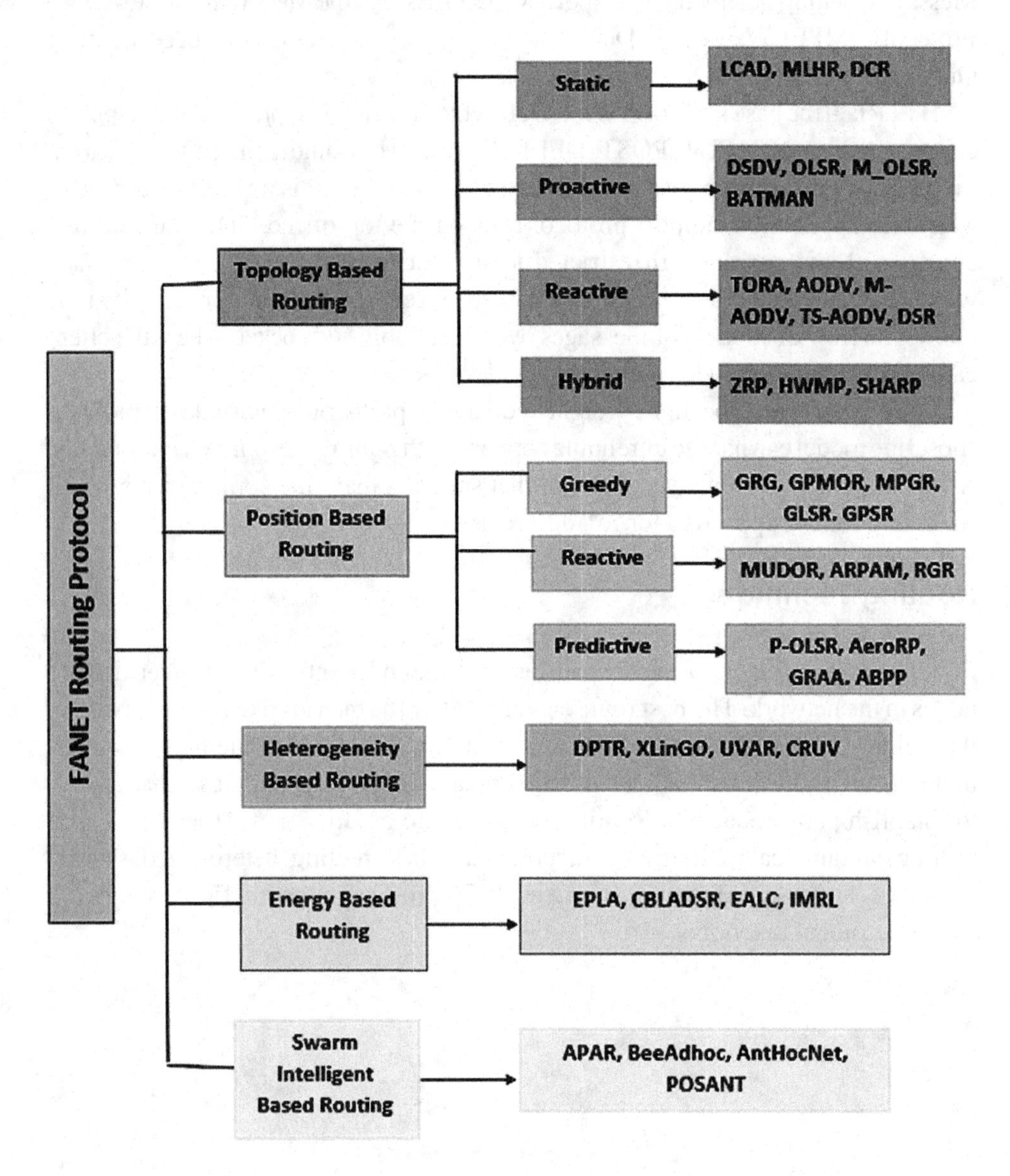


## Topology-based Routing

Topology-based routing is further classified as static, proactive, reactive and hybrid. In static protocol the routing table is fixed throughout the operation, the changes in the node i.e. UAV position cannot be reflected in the table. It holds the various routing protocols like Load Carry and Deliver Routing (LCAD), Data Centric Routing

(DCR), Multi-Level Hierarchical Routing (MLHR). As the FANET dynamically changes its topology this static routing doesn't suit.

The proactive protocol maintains the routing table that holds all the latest updates about the neighbouring UAV nodes. The Optimized Link State Routing (OLSR), Destination Sequenced Distance Vector (DSDV), Mobile-agent based Optimized Link State Routing (MOLSR) and Better Approach To Mobile Ad-hoc Networking (BATMAN) comes under the Proactive protocol. The drawback in this routing is bandwidth not being used efficiently since a lot of data will be exchanged between the nodes for updating the table.

The Reactive routing protocol is an on-demand routing protocol which is a bandwidth efficient protocol. This protocol has Dynamic Source Routing (DSR), Ad-hoc On-demand Distance Vector (AODV), Temporarily Ordered Routing Algorithm (TORA), Multicast Ad-hoc On-demand Distance Vector (MAODV) and Trusted Ad-hoc On-demand Distance Vector (TS-AODV)

Hybrid routing protocol is a combination of reactive and proactive protocols that is suitable to be used in large networks. The participating network is divided into a number of zones, namely, intra-zone and inter-zone. Here the intra-zone routing uses the proactive routing protocol, whereas inter-zone routing uses reactive routing protocol. The Zone Routing Protocol (ZRP), Hybrid Wireless Mesh Protocol (HWMP) and Secured Hierarchical Anonymous Routing Protocol (SHARP) comes under the hybrid routing protocol.

## Energy Based Routing

The FANET nodes are equipped with rechargeable power source which have the capability to recharge continuously during the movement of UAV (Oubbati et al., 2017). The energy-base routing consists of Energy-efficient Packet Load Algorithm (EPLA), Cluster-based Location Aided Dynamic Source Routing (CBLADSR), Energy Aware Link-based Clustering (EALC) and Intelligent Mechanism for Routing data based on nodes Localization (IMRL).

## Swarm Intelligence Based Routing

Swarm intelligence-based routing is adhered from the collective behaviour of the system which doesn't have the decentralized controls and do self-organization. Swarm intelligence-based routing consists of Ant colony optimisation-based Polymorphism-aware Routing algorithm (APAR), BeeAdhoc routing algorithm, Ant Hoc Net, Position-based ANT colony routing (POANT).

## Position Based Routing

Position-based Routing protocol is a geographic-based or location-based routing protocol, which works by the assumption that sender knows about the receiver location and there is no need to maintain the node information (Oubbati et al., 2017). This routing protocol is further classified as greedy, reactive and predictive.

The Greedy routing are highly suitable for dynamic and mobile network. The Greedy random-greedy (GRG), Geographic Position Mobility Oriented Routing (GPMOR), Greedy Perimeter Stateless Routing (GPSR), Mobility Predicted based Geographical Routing (MPGR) and Geographical Load Share Routing (GLSR) are some of the routing protocols comes under greedy routing.

The reactive routing calculates the route only under the need. The Multipath Doppler Routing (MUDOR), Adhoc Routing Protocol for Aeronautical Mobile adhoc networks (ARPAM) and

Reactive-Greedy-Reactive Protocol (RGR) belong to the reactive routing.

The Predictive routing which follows certain predictive approaches, contains the protocols such as Predictive-Optimized Link-State Routing (P-OLSR), Aeronautical Routing Protocol (AeroRP), Geographical Routing Protocol for Aircraft Ad-hoc Network (GRAA) and Adaptive Beacon and Position Prediction (ABPP).

## Heterogeneity Based Routing

Heterogeneity Based Routing support reliable interaction among the UAV nodes and is predicted to be reliable, fast and provide break-free communication. The protocols under this routing are Distributed priority tree-based Routing protocol (DPTR), UAV-aided routing protocol (UVAR) and a Cross-layer Link quality and location aware beaconless opportunistic routing protocol (XLinGo) is a beacon-less routing protocol that support video transmission from one UAV to another.

## Hierarchical Routing

Hierarchical Routing is considered to be the best protocol supported for the network form by UAV. As per the name of this protocol it forms a hierarchical level among the network formed by the UAV nodes, it serves for large size FANET (Sharma et al., 2018). Thus, the FANETs are divided into many clusters and each cluster holds a cluster head (CH) and cluster members nodes. The CH is one among the cluster nodes, which is in the transmission range with all cluster nodes. In general CH takes care about the communicating with upper layers such as satellites and UAVs and it broadcasts data to its cluster members inside the cluster. The two main routing protocols that come under hierarchical routing are Mobility Prediction Clustering (MPC) and Clustering Algorithm (CA) UAV networking.

## AUTHENTICATION

Authentication is an important process in the cyber security that verifies the identity of the user/information. Authentication process in the IoT ecosystem verifies the IoT device's identity when it tries to upload the data to the cloud/server.

The main goal of authentication process is to permit authorized devices/person to access the cloud storage and to rejects the unauthorized users. Usually, in IoT environment, the authentication process is done in two ways,

- Passwords,
- Physical identification, and

### Passwords

Passwords verification is the popular authentication technique used in both computers as well as in the IoT devices. A password is a secret text that may contain alphabets, numbers and symbols. Passwords of the IoT devices are supposed to be known by the user only. Using these passwords, anyone can access the IoT devices and modify the code in it.

### Physical Identification

To improve the security of the IoT devices and cloud storage, physical identification device is used. It includes machine readable cards (i.e.) smart cards, scanning cards etc., In some high security uses cases the password combined with physical identification cards are used. ATM cards is an example of physical identification of authentication process. Some smart cards have an inbuilt username and passwords in the magnetic strips. However, the loss of such smart card can be dangerous.

#### Malicious Node Detection

A specific node in the wireless sensor network or IoT network, that seeks to deny service to other nodes in the network. Also, it may try to breach security principles of the network and may start attacking network. Malicious node may result in some of the following issues to network.

- **Packet Drop**- Simply after receiving the packets, the malicious node may drop the packets without forwarding them.
- **Battery Drained**- A malicious node can enter into the network and waste the battery of other nodes by performing unwanted tasks.
- **Buffer Overflow**- Malicious node enters into the network and pervades the buffer of other IoT devices with a useless value and that may stop the genuine results to be stored or transferred to the cloud storage.
- **Bandwidth Consumption**- A malicious node may consume more bandwidth. So that, the network may face denial of service attack. As a result, no other authenticated/legitimated node can transfer the data packets.
- **Fake Routing-** A malicious node may create a fake path to transfer the data packets to the cloud. So that the legitimate node may send the packet through fake path/route. It may result in unauthorized access and packet loss.

## Machine Learning to Detect the Malicious Nodes in the Network

Machine learning techniques are most commonly used in the detecting the malicious/ suspicious nodes in the wireless sensor network. Machine Learning provides a way of learning strategy without any necessity of being programmed. Just like how human learn by experience, a mechanism where the machine learns which paved a way for the new field of science called Artificial Intelligence.

Machine Learning techniques learns by itself by means of analysing data patterns for malicious node detection. It predicting the unusual behaviour of a node against being malicious and providing defence against suspicious behaviour on the data. Network security solutions are known as Intrusion Detection System (IDS) solutions. Most of the methods proposed are on signature-based approaches. Machine Learning provides newer kind of answer called Network Traffic Analytics (NTA). It aims in proving a deeper exploration of the traffic flow across different levels and finds thrust and suspicious behaviour.

Henceforth, Machine learning came into the picture which is capable of analysing or finding occurrences of an unforeseen event and segregating for further analysis. Machine Learning works in two ways viz testing and training. It helps to analyse and to categorize by using different classifiers, algorithms, etc. One good example of this is analysing the patterns from image data like noses and eyes characteristics thereby resulting in identifying from a large collection of images.

## Certificate Based Access Control

Certificate based authentication is one of the latest techniques used in the IT industries and the government organizations to identify a particular person or entity.

Certificate issuing agency provides a unique digital signature to each customer. With this, the users can enrol the signature for various applications. In some cases, the digital signature might be operated with the tradition username and password method for authentications.

## Working Principle of Certificate Based Access Control

The workflow of the certificate-based access control,

- Initially, the certificate administrator of the organization has to generate and assign the certificate to the individual user or devices.
- Later, the certificate administrator makes the network security systems in the organization to trust the specific user/device
- For logging-on to the network, the user sends a request message to the network security system using the assigned digital certificate.
- The certificate authenticator verifies the digital certificate of the user and allows him to access the network services of the organizations.
- Upon data transferring the message are encrypted and decrypted with they are transferred between the user device and the organizations servers.
- This handshaking process continues between device and server until both are satisfied that the messages each sent have been correctly decrypted and the credentials are sound.
- Mutual authentication ensures that the device is connecting to the server it expects (as only that server could decrypt the message with its credentials) and the server can also verify that the correct device is connecting (as only that device could decrypt the message with its credentials).

## CONCLUSION

FANET is playing a vital role in several fields such as surveillance etc. This chapter focuses on the overall information about FANET, its functionality layers, mobility model and authentication. The necessity of routing protocol concerns with mobility is a major issue needed to be focused by the researchers. The next major issue is related to authentication which protects from an authorised access and it is important for most of real time functionality.

## REFERENCES

Alshbatat, A. I., & Dong, L. (2010). Adaptive MAC Protocol for UAV Communication Networks Using Directional Antennas. *2010 International Conference on Networking, Sensing and Control (ICNSC)*, 598–603. 10.1109/ICNSC.2010.5461589

Bekmezci, I., Sahingoz, O. K., & Temel, Ş. (2013). Flying ad-hoc networks (FANETs): A survey. *Ad Hoc Networks*, *11*(3), 1254–1270. doi:10.1016/j.adhoc.2012.12.004

Bujari, A., Palazzi, C. E., & Ronzani, D. 2017, June. FANET application scenarios and mobility models. In *Proceedings of the 3rd Workshop on Micro Aerial Vehicle Networks, Systems, and Applications* (pp. 43-46). 10.1145/3086439.3086440

Cai, Y., Yu, F. R. R., Li, J., Zhou, Y., & Lamont, L. (2012). *MAC Performance Improvement in UAV Ad-Hoc Networks with Full-Duplex Radios and Multi-Packet Reception Capability*. Advance online publication. doi:10.1109/ICC.2012.6364116

Chriki, A., Touati, H., Snoussi, H., & Kamoun, F. (2019). FANET: Communication, mobility models and security issues. *Computer Networks*, *163*, 106877. doi:10.1016/j.comnet.2019.106877

Ding, Q. (2019). A mathematical model for reflection of electromagnetic wave. *Journal of Physics: Conference Series*, *1213*(4).

Frew, E. W., & Brown, T. X. (2008). Networking issues for small unmanned aircraft systems. *Journal of Intelligent & Robotic Systems*, *54*(1), 21–37.

Gozalvez, J., Sepulcre, M., & Bauza, R. (2012). Impact of the radio channel modelling on the performance of VANET communication protocols. *Telecommunication Systems*, *50*(3), 149–167. doi:10.100711235-010-9396-x

Jasim, M. M., Al-Qaysi, H. K., Allbadi, Y., & Al-Azzawi, H. M. (2020). Comprehensive study on unmanned aerial vehicles (UAVs). Advanced Mathematical Models & Applications, 5(2), 240-259.

Khan, I. U., Qureshi, I. M., Aziz, M. A., Cheema, T. A., & Shah, S. B. H. (2020). Smart IoT control-based nature inspired energy efficient routing protocol for flying ad hoc network (FANET). *IEEE Access: Practical Innovations, Open Solutions*, *8*, 56371–56378. doi:10.1109/ACCESS.2020.2981531

Khan, M. A., Qureshi, I. M., & Khanzada, F. (2019). A hybrid communication scheme for efficient and low-cost deployment of future flying ad-hoc network (FANET). *Drones (Basel)*, *3*(1), 16. doi:10.3390/drones3010016

Kheli, F., Bradai, A., Singh, K., & Atri, M. (2018). Localization and Energy-Efficient Data Routing for Unmanned Aerial Vehicles: Fuzzy-Logic-Based Approach. *IEEE Communications Magazine*, *56*(4), 129–133. doi:10.1109/MCOM.2018.1700453

Lakew, D. S., Sa'ad, U., Dao, N. N., Na, W., & Cho, S. (2020). Routing in flying ad hoc networks: A comprehensive survey. *IEEE Communications Surveys and Tutorials*, *22*(2), 1071–1120. doi:10.1109/COMST.2020.2982452

Lin, N., Gao, F., Zhao, L., Al-Dubai, A., & Tan, Z. (2019, August). A 3D smooth random walk mobility model for FANETs. In *2019 IEEE 21st International Conference on High Performance Computing and Communications; IEEE 17th International Conference on Smart City; IEEE 5th International Conference on Data Science and Systems (HPCC/SmartCity/DSS)* (pp. 460-467). IEEE.

Oubbati, O. S., Lakas, A., Zhou, F., Güneş, M., & Yagoubi, M. B. (2017). A survey on position-based routing protocols for Flying Ad hoc Networks (FANETs). *Vehicular Communications*, *10*, 29–56. doi:10.1016/j.vehcom.2017.10.003

Sawalmeh, A. H., & Othman, N. S. (2021). *An overview of collision avoidance approaches and network architecture of unmanned aerial vehicles (UAVs).* arXiv preprint arXiv:2103.14497.

Sharma, V., Kumar, R., & Kumar, N. (2018). DPTR: Distributed priority tree-based routing protocol for FANETs. *Computer Communications*, *122*, 129–151. doi:10.1016/j.comcom.2018.03.002

Temel & Bekmezci. (2015). LODMAC: Location Oriented Directional MAC Protocol for FANETs. *Computer Networks, 83*, 76–84. . doi:10.1016/j.comnet.2015.03.001

Tran, T. X., Hajisami, A., & Pompili, D. (2017). Cooperative Hierarchical Caching in 5G Cloud Radio Access Networks. *IEEE Network*, *31*(4), 35–41. doi:10.1109/MNET.2017.1600307

Vasilyev, G. S., Surzhik, D. I., Kuzichkin, O. R., & Kurilov, I. A. (2020). Algorithms for Adapting Communication Protocols of Fanet Networks. *Journal of Software*, *15*(4), 114–122. doi:10.17706/jsw.15.4.114-122

Wang, R., Taleb, T., Jamalipour, A., & Sun, B. (2009). Protocols for reliable data transport in space internet. *IEEE Communications Surveys and Tutorials*, *11*(2), 21–32. doi:10.1109/SURV.2009.090203

# Chapter 5
# A Review of Various Modeling Software for VANETs:
## Simulation and Emulation Tools

**Divya L.**
*Pondicherry Engineering College, India*

**Pradeep Kumar T. S.**
https://orcid.org/0000-0001-7071-4752
*Vellore Institute of Technology, Chennai, India*

## ABSTRACT

*There is great demand for VANETs in recent times. VANETs enable vehicular communication with the advent of latest trends in communication like 5G technology, software-defined networks, and fog and edge computing. Novel applications are evolving in recent times on VANETs with the proliferation of internet of things. Real test bed implementation is not always feasible with various limitations like expenditure and manpower and requires more time to experiment with the new facets of VANETs. Hence, the researchers should be aware of the variety of simulation tools that are capable of running VANET simulations. Simulation is a powerful tool in developing any critical/complex system that constitutes minimum cost and effort. The simulation tools of VANETs should support multiple mobility models, real-world communication protocols, and traffic modeling scenarios. This chapter gives a clear view on available tools and their characteristics on VANETs for research purposes.*

DOI: 10.4018/978-1-6684-3610-3.ch005

## INTRODUCTION

The vehicular mobility in VANETs relies on micro and macro mobility. Communication protocols support interchanging data between vehicles and RSUs.

VANETs simulation is possible in all layers of Open System Interconnection (OSI) reference model. Agarwal et al., developed a TraceReplay simulator at application layer to implement realistic implementation of applications in NS3. TraceReplay works from the network trace information to generate real world applications in NS3. Any application layer protocols like Hyper Text Transfer Protocol (HTTP) can be replayed with this TraceReplay proposed by Agrawal (2016). Jang (2017) proposed overlay platform at application layer with Greedy Perimeter Stateless Routing (GPSR) and Ad-hoc On-demand Distance Vector (AODV) protocols to increase reliability in VANET communication. Monir (2022) proposed seamless Mobile Edge Computing (MEC) based SDN handover management scheme for VANETs for handling the mobility challenges in VANETs. MEC server at Road Side Unit (RSU) runs the handover logic upon the intersection of adjacent RSU's. The handover is done when RSSI values falls below the defined threshold. The limitation of the proposed scheme is it is unable to handle cross roads handover. Summarized list of existing works is given in Table 1.

*Table 1. Existing Works*

| Defined Protocol | Software Used | Protocol Used | Mobility Tool | OSI Layer |
|---|---|---|---|---|
| TraceReplay (Agrawal, 2016) | NS3 | PCAP Trace, TCP | SUMO | Application |
| Overlay-G, Overlay-A (Jang, 2017) | NS2 | GPSR, AODV | MOVE, SUMO | Application |
| MEC/ SDN VANET (Monir,2022) | NS3 | Open Flow switch | SUMO | Cross Layer |
| CPMRA & CPDRA (Deeksha et al, 2022) | NS3 | DSRC, Cellular communication | SUMO | Transport |
| Adaptive Jumping Multi-Objective Firefly Algorithm (AJ-MOFA) (Hamdi et al, 2022) | MATLAB 2019b | FIREFLY algorithm | NA | Network |
| IoDAV (Ahmed et al, 2021) | VEINS, OMNET++ | Improved Particle Swarm Optimization (IPSO) | SUMO NetEdit, SUMO | Cross layer |
| LRF (Din et al, 2020) | NS3 | Information Centric Networking (ICN) | SUMO | Network level |
| RGoV (Kazi et al, 2021) | NS2 | K means | SUMO | Network |
| BCOOL (Maaroufi et al, 2021) | NS 3.26 | Open Flow protocol, Balogna data set, Broad trip protocol, IEEE 802.11 MAC | SUMO | Cross layer |
| MDPRP (Aznar et al, 2021) | Omnet++ | IEEE 802.11p | INET 3.5 LIBRARY | Transport |
| PHBMAC (Yang et al, 2022) | MAT lab | Markov model, Information priority | ---- | MAC |

Deeksha et al, (2022) proposed a decentralized congestion control algorithm based on power, data and message rates using the state transitions specified by European Telecommunications Standards Institute (ETSI). The authors proposed two algorithms Combined power and data rate adaption (CPSRA) and Combined Power and Message Rate Adaption (CPMRA) and tested over real world scenarios generated by SUMO. The authors enhanced transmission channel utilization by limiting the congestion. The limitation of the work fair allocation of channels should be considered. Hamidi et al, (2022) developed Adaptive Jumping Multi Objective Firefly optimization (AJMOFA) algorithm. This algorithm optimizes the data distribution, network performance in VANETs with clustering, data forwarding by considering the factors like packet drop, delay and delivery rate. This algorithm in computation intensive and may not be suitable for emergency situation in VANETs. Ahmed et al, (2021) developed a technique called Internet of Drones Assisted VANETs (IoDAV) to increase the coverage area of IoD with the consideration of vehicle location. IoDAV handles dynamic mobility of vehicles using the RSSI factor.

Din et al., proposed Left Right Front (LRF) caching technique in collaboration with ICN. This technique selects best node to cache data for preserving emergency data without data loss between RSUs. This technique can be enhanced for infotainment and V2V communications. Kazi et al, (2021) proposed clustering based routing protocol named Routing Group of Vehicles (RGoV) to avoid unnecessary broadcasting of data to unreliable nodes. The node clusters were formed with reliable nodes with restricted communication area and K means algorithm to minimize delay and maximize packet delivery ratio. This protocol considered only few network parameters and parameters like RSSI, packet drop can also be considered. Maaroufi et al, (2021) developed Blockchain COngestion ContrOL (BCOOL) architecture with trust, congestion and virtual network function placement modules. This novel architecture can handle connectivity failures and maintain QoS along with security in VANETs . Juan et al., anticipated Markov Decision Process Rate and Power (MDPRP) distributed model to limit channel load with beacon data packets in the network to reduce congestion in the network . Yang et al, (2022) developed a Priority Based Hybrid Medium Access Control protocol (PHBMAC) that categorizes messages and channel contention. This protocol used Unmanned RSUs for geographical coverage fo vehicles in the VANET. TDMA is implemented for channel contention between RSUs to RSus and OBUs to RSUs .

## LIST OF MODELING AND EMULATION TOOLS

### Vehicular NeTwork Open Simulator (VENTOS)

VENTOS is an open source C++ based simulation tool working with Dedicated Short Range Communication (DSRC) for VANET communication. VENTOS works in combination of SUMO and OMNET++ for traffic generation and network communication. Researchers and developer community use VENTOS for vehicular communication like Vehicles to Vehicles (V2V), Vehicle to Infrastructure (V2I), and analysing flow of traffic for traffic management. It was developed by rubinet lab. PLATOON management protocol and Traffic Signal Control (TSC) in standalone intersection are supported by VENTOS. This simulation tool provides adaptive routing of traffic algorithms based on real traffic to minimize the delay.

## NS3

This is free, open source software under GNU GPLV2 license. NS3 acts as a simulator and emulator tool that collaborates with real world protocols. C++ and python languages are used to code in NS3. This tool enables to create virtual nodes that behave as actual nodes, supports P2P, CSMA, wireless technologies. This also support advanced technologies like SDN with open flow switches and network interface card for network connectivity. Network metrics like data rate, QoS, PDR, packet size can be quantifies with this tool. It allows to generate .plt files with GNUPLOT and pcap files (wireshark), NetAnim for visualization of actual network and node trace to know about transmission and reception of data ($T_x$ and $R_{x).}$ NS3 allows integrating SUMO to model real world traffic with desired geographic locations for simulation.NS3 controls vehicle state in SUMO. The position of vehicles is synchronized with the NS3 nodes. The limitation of SUMO is all the nodes need to be preconfigured in NODEPOOL before starting the simulation. NS3 software has incremental releases twice or thrice a year after updates and after bug fixes in the existing versions.

## Objective Modular NETwork (OMNET++)

This is open source, modular, object oriented, simulation tool for modeling VANETs. This is programmed in C++. OMNET++ supports wired and wireless protocols like PPP, ethernet, STP, RSTP, TDM/WDM, IEEE 802.11 p/e, IEEE 802.15.4, IEEE 802.16 e (WiMAX). OMNET ++ at application layer supports HTTP, file transfer with INET framework and dynamic host configuration protocol, basic video streaming and network attack (NETA) OMNET ++ also supports mobility environment with VEINS (Vehicle In Network Simulator) for vehicular mobility and satellite (OS). OMNET ++ has kernel for simulation and network functionality like statistics, characterization with utility classes. OMNET++ modules are declared in (Network Description (NED) language. Network emulation is possible with OMNET++ along with simulation with event scheduler and hardware in the loop functionality. Node communication in OMNET++ is provided by INET packet package. A network in OMNET++ is formed with the interconnection of modules. OMNET++ is extensible.

## JSIM

JSIM is similar to OMNET but developed in JAVA. Assembling of modules in JSIM is similar to NS2. JSIM works in collaboration with JAVA & Tool Command Language (TCL). Running the TCL scripts is a problem in JSIM as graphical editors are not feasible to run the TCL files.. JSIM has graphical Editor that can export TCL files, XML are loaded straight into the simulator without TCL. But, XML files are complex and are not easy to understand. JSIM supports IPV4, TCP, MPLS. JSIM enables faster development than C++ but experiences low performance. JSIM cannot run real world protocols like OMNET++ and NS3.

## QUALNET

This is a commercial simulation tool for simulating network security issues. QUALNET is suitable for simulating Adhoc, wired, wireless networks. Network attacks like eaves dropping, radio signal jamming, Denial of Service (DOS) attacks and Distributed Denial of Service (DDOS) attacks can be simulated. QUALNET has an animator, designer for protocol design, analyzer for statistical analysis and tracer for packet tracing.

## Vehicles in Network Simulation (VEINS)

An open source simulation tool to generate vehicular traffic (network) scenarios that was used across five continents. It is based on INET, SUMO mobility model and OMNET++ network simulator for rapid network setup and executing interactive simulations. VEINS provide inter vehicle communication in vehicular network. Many automobile industries, government organizations and research teams were working on VEINS for traffic analysis. The various features of VEINS include 3 –D visualization, re-routing in case of traffic jams and accidents. CEINS also permits reconfiguring of vehicles, depends on trusted vehicular model for mobility, enables OSM scenarios, supports IEEE 802.11P and IEEE 1609.4 DSRC/WAVE at network layer. VEINS allow using various metrics like emissions and time to travel in traffic simulations.

### VEINS Architecture

The architecture of VEINS is shown in Fig1. The network and mobility tools were bi-directionally connected over TCP socket. OMNET++ provides simulation control of discrete events and generates the results. Road traffic is generated by SUMO and VEINS enables traffic efficiency, safety and comfort. Traffic Control Interface (TraCI) is the communication protocol in VEINS that acts as an interface between OMNET++ and SUMO. The vehicles in SUMO were interpreted as nodes in OMNET++.

*Figure 1. VEINS architecture*

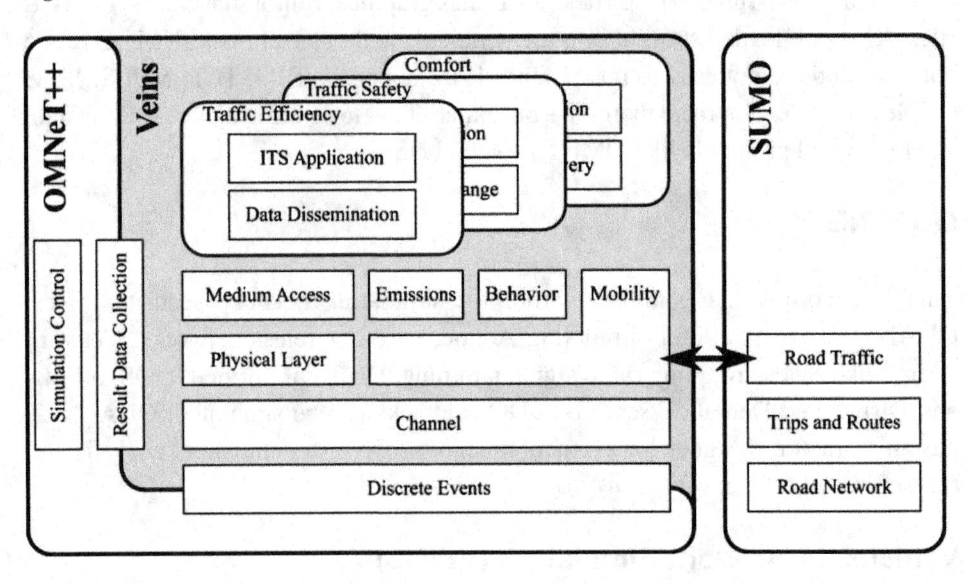


## NETwork SIMulator (NETSIM)

NETSIM is commercial full stack, host to host, packet level simulation tool to simulate and emulate various types of networks developed by TETCOS. NETSIM is available as NETSIM professional, standard and academic. NETSIM pro is designed for enterprise and defense purposes, on the other hand standard and academic focus on educational clients like researchers and student community. Packet animator in NETSIM animates the flow of packets in the network. Network performance metrics can be analyzed in graphs and plots at multiple levels in the whole network. NETSIM provides interface with SUMO shown in Fig 2. and MATLAB software's. NETSIM supports researchers to develop and modify new and existing networking protocols.

*Figure 2. Working of NETSIM*

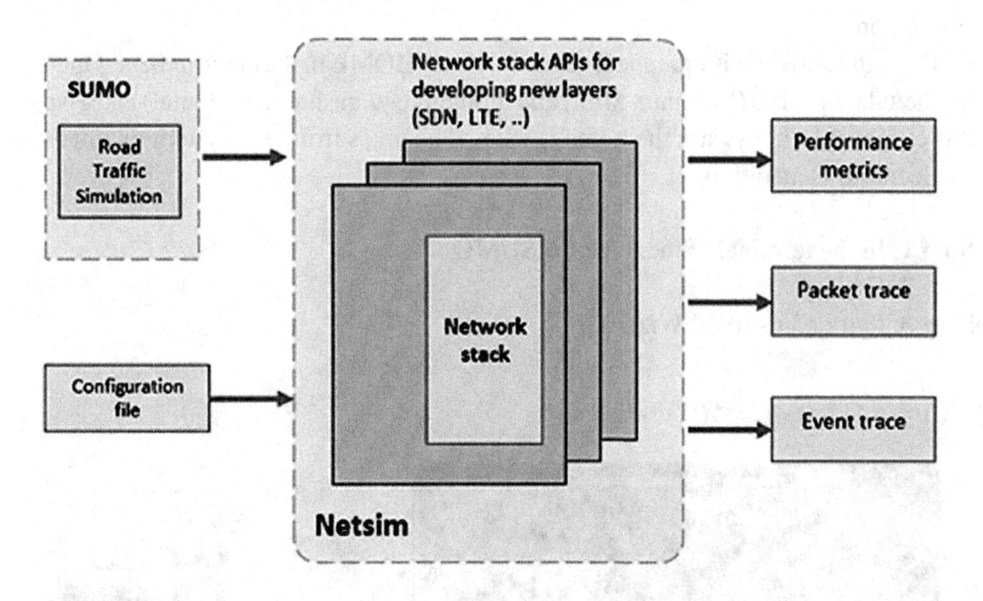


## Basic Traffic Simulation in SUMO

Simulation of Urban MObility (SUMO) is a mobility modeling tool to generate urban traffic for modeling VANET scenarios. SUMO is compatible with many network simulation tools OMNET++, NS3, NS2. SUMO permits the user to select any geographical location throughout the world in .osm an xml based file with www.openstreetmap.org. Sumo converts the .osm file to its native xml file to be interoperable & portable with network simulation tools for animating and tracing the network. The creation of real traffic network in SUMO includes the following steps:

### Steps to Install and Scenario Generation in SUMO on Ubuntu 20.04s

Prerequisites:

**Step 1:** sudo add-apt-repository ppa:sumo/stable
**Step 2:** sudo apt-get update
**Step 3:** repeat step 1.

Also, Download the sumo source file from https://sumo.dlr.de/releases/1.12.0/sumo-src-1.12.0.tar.gz and unzip or untar it to the home directory (/home/uday). There are some python files that are needed to generate random trips and to export xml files to tcl files. The commands sumo, sumo-gui will run only the graphical simulation.

Once the software is unzipped, set the SUMO_HOME in the environment, Open /home/uday/.profile (for Linux Mint) and /home/uday/.bashrc (for Ubuntu) and give this command (in my case the home is uday, it changes from one system to other)

Scenario Generation:

**Step 1**: In the terminal home, type Cd SUMO
**Step 2**: Cd tools
**Step 3**: Python3 osmWebWizard.py

*Figure 3. Opening OSMWebWizard.py*

It opens a tab in the web browser with a map of Berlin as shown in Figure 4.

*Figure 4. OSMWebWizard by default position Berlin*

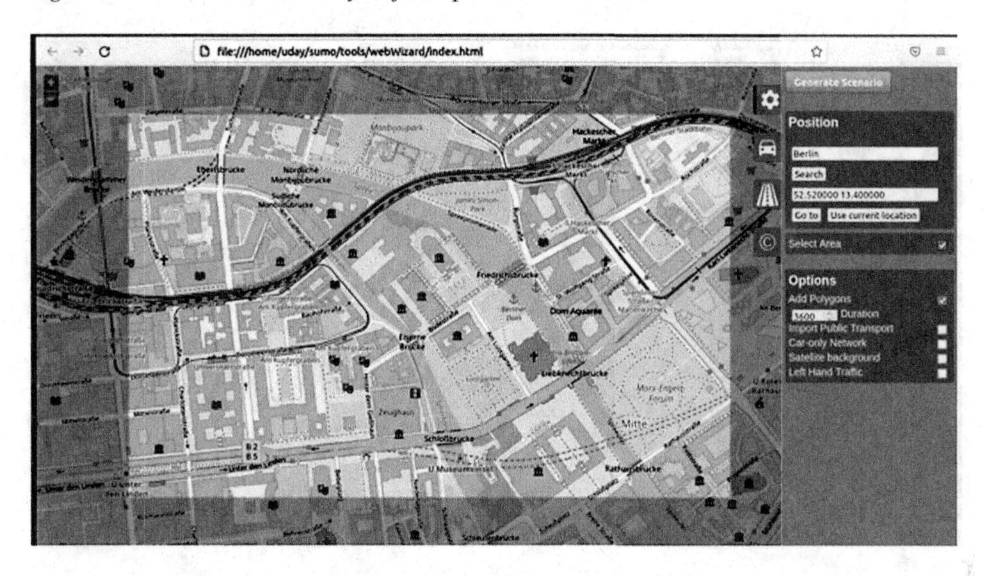


**Step 4:** Now tick check box Select Area and select the area you want to generate (Here I have selected Pondicherry, India) as depicted in Fig. 5.
**Step 5:** Select the wizard options from the list given below

By default the "Add Polygon" checkbox is checked and a road traffic simulation is generated but all types of roads and rails will be imported as well (cycle paths, footpaths, railways etc) as shown in Fig. 5.

- Enable checkbox "left-hand Traffic" for left hand traffic rules.
- Enable "Car-only Network" to permit passenger car traffic.
- Enable"Import Public Transport" checkbox to enable bus and train stops.
- Enable Demand-checkbox "Bicycles" checkbox, then extra bicycle lanes will be added.
- Enable Demand-checkbox "Pedestrians" checkbox, then OSM generates sidewalks and pedestrian crossings.

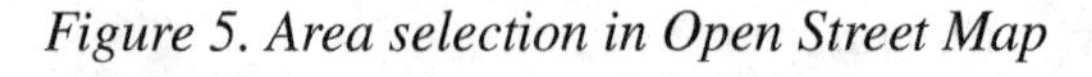

*Figure 5. Area selection in Open Street Map*

**Step 6:** SUMO supports various transportation modes; those modes are included in the OSM for traffic generation. Transportation modes are selected by clicking the respective checkboxes.

These modes are operated on demand by probability distribution influenced by parameters like COUNT and Through Traffic factor.

- *Through Traffic Factor* defines how many times it is more likely for an edge at the boundary of the simulation area being chosen compared to an edge entirely located inside the simulation area.
- The vehicles generated per hour are the *Count* parameter.

**Step 7:** Road type selection provides highway, pedestrian, railway, aero way, water way.

**Step 8:** Click on Generate Scenario that pops up SUMO-GUI window. Then run the simulation where we can see the vehicles moving on the open street map as shown in Fig. 6.

*Figure 6. SUMO GUI after scenario generation*

## Traffic Software Integrated System – CORridor SIMulation (TSIS-CORSIM)

MCTRANS consists of three packages HCS (Highway Capacity Software), Highway Safety Software (HSS), TSIS-CORSIM. HCS allows macroscopic modeling of surface street facilities and connecting automated vehicles. HSS evaluated prediction of crashes on roads by using Highway Safety Manual (HSM). TSIS CORSIM is a microscopic commercial mobility simulation model. TSIS- CORSIM is a hybrid version that was developed from NETwork SIMulation (NETSIM) and FREeway SIMulator (FRESIM). Urban traffic is simulated with NETSIM and FRESIM simulates traffic on highways and freeways. Traffic can be simulated by driver behavior and individual vehicles. U.S department of transportation Federal HighWay Administration (FHWA) used CORSIM for research on traffic control systems, .

## City Flow

An open source multi agent reinforcement learning based simulator with interactive user interface. City flow supports vehicular network creation and flow of traffic on real time and artificial traffic. City flow provides controlling traffic signals and traffic analysis. City Flow allows creating large traffic scenarios with more intersections and vehicles on roads. The various features of City Flow are efficient, scalable, reproducible and effective simulation tool compared to SUMO .

## Comparison Statement of Various Modelling Tools

Various VANET modelling tools are contrasted and compared in the table 2 given below.

*Table 2. Comparison of various vanet modelling*

| Tool | Language | Medium of Complexity | Software Kind | Configuration Settings | Disadvantages |
|---|---|---|---|---|---|
| NS3 | C++ and Python | Moderate | Open | GNU/Linux, Windows, FREE BSD and Mac OS | Difficult to model real system |
| VEINS | C++ | Moderate | Open | Windows, Linux, Mac OS | Needs both OMNET++ and SUMO |
| OMNET++ | C++ | Moderate | Open | Windows, Linux, Mac OS | Compatibility issues and available protocols are less. |
| TOSSIM (Al-Roubaiey, A., & Al-Jamimi 2019) | Python and C++ | Moderate & Obsolete | Open | Linux, Cygwin on Windows | Problem when running on real mote. |
| NS2 | TCL and C++ | Complex | Open | LINUX, Windows, FREE BSD and Mac OS | Poor GUI, needs grip on TCL |
| VENTOS (Amoozadeh, 2019) | C++ | Complex | Open | Windows, Linux, Mac OS | Both SUMO and OMNET++ |
| J SIM | JAVA and TCL | Complex | Open | Windows, Linux, Mac OS | Needs more time for execution |
| NETSIM | JAVA | Moderate / Proprietary | NETSIM STD is open source, others are Commercial | Windows, Linux, Mac OS | Lack of thread synchronization. Download times varies timely. |

## CONCLUSION

This paper handles various modeling and simulation software's for VANETs. Most of them are free and open source software's. We have done a basic simulation in urban mobility and we have showed how to go for real word traffic modeling using Open Street Maps (OSM). Students and researchers can use these tools for research purpose and affording a vehicle for getting values on road is a complex and costly task.

So this chapter can be used by researchers and students to know various simulation and emulation tools for developing various novel applications.

## REFERENCES


Agrawal, P., & Vutukuru, M. (2016). Trace based application layer modeling in ns-3. *2016 Twenty Second National Conference on Communication (NCC)*, 1-6. 10.1109/NCC.2016.7561126

Ahmed, G. A., Sheltami, T. R., Mahmoud, A. S., Imran, M., & Shoaib, M. (2021). A novel collaborative IoD-assisted VANET approach for coverage area maximization. *IEEE Access: Practical Innovations, Open Solutions*, *9*, 61211–61223. doi:10.1109/ACCESS.2021.3072431

Al-Roubaiey, A., & Al-Jamimi, H. (2019, June). Online power Tossim simulator for wireless sensor networks. *2019 11th International Conference on Electronics, Computers and Artificial Intelligence (ECAI)*, 1-5. 10.1109/ECAI46879.2019.9042005

Amoozadeh, M., Ching, B., Chuah, C. N., Ghosal, D., & Zhang, H. M. (2019). VENTOS: Vehicular network open simulator with hardware-in-the-loop support. *Procedia Computer Science*, *151*, 61–68. doi:10.1016/j.procs.2019.04.012

Aznar-Poveda, J., Garcia-Sanchez, A. J., Egea-Lopez, E., & Garcia-Haro, J. (2021). Mdprp: A q-learning approach for the joint control of beaconing rate and transmission power in vanets. *IEEE Access: Practical Innovations, Open Solutions*, *9*, 10166–10178. doi:10.1109/ACCESS.2021.3050625

Corridor simulation (CORSIM/TSIS). (n.d.). *Traffic Analysis Tools: Corridor Simulation - FHWA Operations*. https://ops.fhwa.dot.gov/trafficanalysistools/corsim.htm

Deeksha, M., Patil, A., Kulkarni, M., Shet, N. S. V., & Muthuchidambaranathan, P. (2022). Multistate Active Combined Power and Message/Data Rate Adaptive Decentralized Congestion Control Mechanisms for Vehicular Ad Hoc Networks. *Journal of Physics: Conference Series*, *161*(1), 012018. doi:10.1088/1742-6596/2161/1/012018

Din, I. U., Ahmad, B., Almogren, A., Almajed, H., Mohiuddin, I., & Rodrigues, J. J. (2020). Left-right-front caching strategy for vehicular networks in icn-based internet of things. *IEEE Access: Practical Innovations, Open Solutions*, *9*, 595–605. doi:10.1109/ACCESS.2020.3046887

Hamdi, M. M., Audah, L., & Rashid, S. A. (2022). Data dissemination in VANETs using clustering and probabilistic forwarding based on adaptive jumping multi-objective firefly optimization. *IEEE Access: Practical Innovations, Open Solutions, 10*, 14624–14642. doi:10.1109/ACCESS.2022.3147498

Jang, J., Ahn, T., & Han, J. (2017). A new application-layer overlay platform for better connected vehicles. *International Journal of Distributed Sensor Networks, 13*(11). Advance online publication. doi:10.1177/1550147717742072

Kazi, A. K., Khan, S. M., & Haider, N. G. (2021). Reliable group of vehicles (RGoV) in VANET. *IEEE Access: Practical Innovations, Open Solutions, 9*, 111407–111416. doi:10.1109/ACCESS.2021.3102216

Maaroufi, S., & Pierre, S. (2021). BCOOL: A novel blockchain congestion control architecture using dynamic service function chaining and machine learning for next generation vehicular networks. *IEEE Access: Practical Innovations, Open Solutions, 9*, 53096–53122. doi:10.1109/ACCESS.2021.3070023

McTrans. (2022, August 18). *McTrans center*. https://mctrans.ce.ufl.edu/

Monir, N., Toraya, M. M., Vladyko, A., Muthanna, A., Torad, M. A., El-Samie, F. E. A., & Ateya, A. A. (2022). Seamless Handover Scheme for MEC/SDN-Based Vehicular Networks. *Journal of Sensor and Actuator Networks, 11*(1), 9. doi:10.3390/jsan11010009

Tetcos. (n.d.). *Tetcos: NetSim - Network Simulation Software, India*. https://www.tetcos.com/

Windows#. (n.d.). *Installing - SUMO Documentation*. https://sumo.dlr.de/docs/Installing/index.html

Yang, X., Mao, Y., Xu, Q., & Wang, L. (2022). Priority-Based Hybrid MAC Protocol for VANET with UAV-Enabled Roadside Units. *Wireless Communications and Mobile Computing, 2022*, 1–13. doi:10.1155/2022/8697248

# Chapter 6
# Real-Time Traffic Simulation of Vehicular Ad hoc Networks

**Pradeep Kumar T. S.**
https://orcid.org/0000-0001-7071-4752
*Vellore Institute of Technology, Chennai, India*

**Vetrivelan P.**
*Vellore Institute of Technology, Chennai, India*

## ABSTRACT

*Vehicular ad hoc networks (VANETs) are a type of ad hoc networks where the node movements are high and there will be instant communication between the vehicles (nodes). In this chapter, the authors propose a real-time simulation of vehicular ad hoc networks using simulation of urban mobility (SUMO) in two cases: 1) user-defined road structure and 2) roads designed through open street maps. In both these cases, cars, buses, trucks, pedestrians, and bicycles will be running in the roads. Most of the vehicles will be following the Euro emission norms. Later these cars will be modelled as nodes in a network and analyse the various network performance metrics like throughput and packet delivery ratio were computed.*

## INTRODUCTION

Vehicular Adhoc networks or VANETs play a major role for various automotive companies for their infotainment systems, car to car connectivity, connecting to Roadside Units (RSU) in case of emergencies like accidents, forest fire, etc. In this chapter, a real world simulation is being carried out using the software Simulation of Urban Mobility (SUMO) for different road design. One of the design would be

DOI: 10.4018/978-1-6684-3610-3.ch006

to design a road by any user based on conditions like type of driving (Right Hand or Left Hand), lanes in each road, etc. The other design would be select a Map in the globe through open street map and simulate the network. Many works have been carried out by the researchers and students across the globe in this area. Some of those works are listed here.

Wang, J., et al., (2020) proposed a context aware quantification method for VANETs using Markov Chain process. The simulation is carried out analytically with a creation of a state transition matrix through discrete Markov Chain and Poison process. The security strategies were imitated for the experimentation. Seliem, H et al., (2018) proposed a drone with vehicular communication to minimise the delay between the drones and the vehicles on road. The simulation is carried out in a two-way highway. This paper used a method called as the Drone active service which will identify the location of the vehicles on road. Their results prove that the packet delivery delay between the drones and the vehicles are minimum.

Liu, B et al., (2015) suggested a method for safety messages between the vehicles. In this paper, cloud provisioning is used to deliver the safety message initially to the gateways. Through the gateways, the messages will be delivered to the vehicles nearby. Later using V2V technology, all the vehicles in the networks can get the safety messages. Since its cloud assisted platform, there could some latency in delivering the messages. Alharthi, A., et al., (2021) proposed a blockchain based biometric framework for vehicular networks to store the biometric data secure in a block chain to prevent attacks like Sybil and replay attacks. The simulations were carried out in OMNet++ along with VEINS and calculated the PDR, PLR and the computational cost of such a network. Cheng, C. M., & Tsao, S. L. (2014) implemented a new protocol and simulated using SUMO and QualNet. This protocol deals with a bloom filter that improves the efficiency of the information retrieval system while using the two tier architecture namely VANET and P2P. This paper reduces the latency and overhead by 12 and 20% respectively.

Kumbhar, F. H., & Shin, S. Y. (2020) proposed a fog based node architecture that can predict the most suitable path based on the machine learning models adopted for the VANETs. This work is simulated using SUMO which also measures the packet delivery ratio and the connectivity of up to 4 hops. Haghighi, M. S., & Aziminejad, Z., (2019) have implemented an onion based routing protocol that can maintain anonymity for the source, destination and even the route. The simulation is carried out using SUMO and its compared against the naïve onion protocol that compares with delivery ratio, retransmission and end to end delay. Abuashour, A., & Kadoch, M., (2017) introduced three cluster based routing protocols that can improvise the parameters like delay, throughput, route stability and control overhead messages. The traffic generation is carried out in SUMO and compared it against MATLAB.

Shah, A. S., (2020) proposed a cluster based Medium Access Control protocol that can variably increases or decreases the number of clusters in VANETs according to the number of vehicles in the network. In this work, the transmission probability is optimized due to the change is number of clusters. The mobility models for the work is simulated using SUMO and measure performances like throughput, delay and PDR. This work achieves higher throughput while there is a shortfall in the PDR and delay. Contreras, M., & Gamess, E. (2020) proposed algorithm that find the number of vehicles in a traffic signal by making any one vehicle as a region leader. That leader is responsible for sending or communicating any important or safety messages to other leaders or vehicles in the network. Also the leaders sometime act as routers as well. The entire simulation is carried out using VEINS, SUMO, OMNetT++ for the traffic generation and network simulation. Also TCP is the transport layer used in this work.

Luo, G., (2018) proposed a MAC protocol based on Software Defined networks with the road side units (RSU) were modelled as an open flow switch which is controlled by a SDN controller. Also Tiennoy, S., & Saivichit, C. (2018) proposed a distributed road side unit that caters to the distributed architecture of named network. This work proposed a RSU model that can disseminate data between then RSUs and interact with the vehicles on the road. The prime objective of this work is to separate the control and data part so that the network safety is ensured and rapid mobility is increased due to the varying number of vehicles in the network. Ahmad, F., (2018) and Akhtar, (2014) proposed a trust evaluation and management framework to test the various models of trust for the evaluation of vehicles for its malicious behaviour. The framework is simulated using OMNet++ and SUMO for the network simulation and traffic generation. These kinds of frameworks are very helpful for the car manufacturers and for planning smart cities. Kumar, T. P., & Krishna, P. V. (2021) surveys various techniques, simulators and strategies for internet of vehicles and how the simulation can be performed and how well they can be modelled for energy efficiency. Most of the literature collected above uses SUMO, OMNeT++, ns3 or ns2 as simulation tools and used open source tools for the analysis of vehicular Adhoc networks.

## TOOLS USED FOR REAL WORLD SIMULATION OF VANETS

Vehicular Adhoc networks can be simulated or emulated using many different tools and software. In this chapter, we will be discussing about the tools which will be focussed here. The Figure 1 shows the complete process of generation of traffic pattern from SUMO to NS2 or NS3.

- Design of traffic model using SUMO with open street maps or design of own road structure.
- Generation of configuration files related to SUMO
- From the config files, mobility config is generated that shows the movement of vehicles as nodes.
- The mobility file is used to program in to TCL (Tool Command Language) for generating the simulation for either ns2 or ns3.
- Once the script of ns2 or ns3 is run, an animation and set of trace or log files is generated for further analysis.

*Figure 1. Process of Generation of Traffic in SUMO*

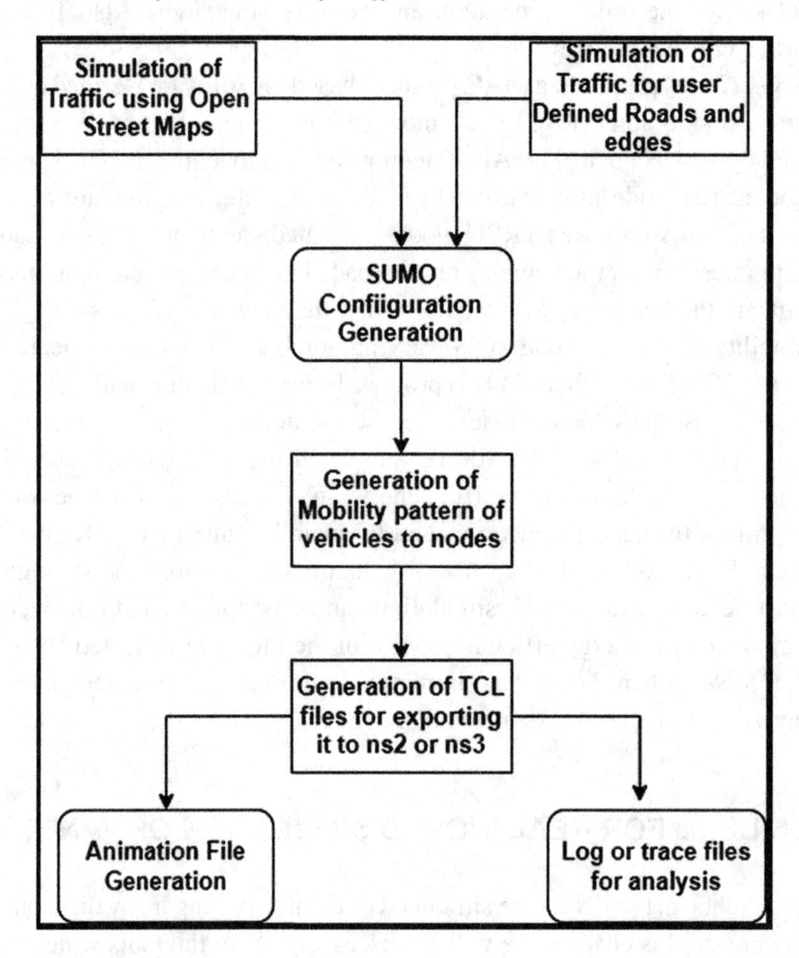

## SUMO (Simulation of Urban Mobility)

Simulation of Urban Mobility (SUMO) is one of a powerful tool where the microscopic traffic generation is possible. There will be floating cars in the GUI of SUMO which shows many vehicles namely cars, buses, trucks, motor cycles, bicycles, pedestrian, etc. SUMO is a open source tool and it is easily portable and can handle a very large network.

## Network Simulator 3

ns-3 is a discrete-event network simulator for Internet systems, primarily for research and educational use. ns-3 is free and open-source software. NS3 can simulate networks like wired, wireless systems, sensor networks, satellite networks, all layers' protocol simulation, etc. NS3 generate log or trace files of ascii traces and also generates pcap files.

## Open Street Maps (OSM)

Open Street Maps is a free and open-source tool that can generate OSM files that can be imported to many applications that support osm. Its equivalent to Google Maps API but it's completely free. Open street map can generate a scenario generation in just 3 clicks.

# EXPERIMENTAL RESULTS AND ANALYSIS

In this chapter, there will be simulations of two types,

- OSM based Simulation
- User Defined road based simulation.

## OSM Based Simulation

Open Street Maps (OSM) generates an XML file with a file extension of .osm which contains information about the various places on the earth based on their latitude and longitude. In the osmWebwizard.py file, the <latitude, longitude> along with the type of vehicles can be mentioned. Also a rectangular area can be set as indicated in Figure 2. From the figure, either the latitude and longitude can be mentioned or else the name of the place can be mentioned. Also there is an option called "Select Area" which can select a random rectangular area for which the simulation can be performed. The duration of simulation is also set as 600 seconds and the type of traffic is selected as “Left hand traffic” so that it will be suitable for countries like India and some parts of Europe.

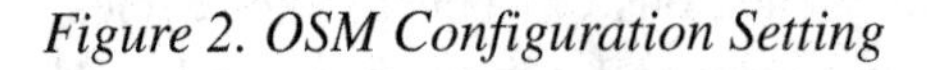

*Figure 2. OSM Configuration Setting*

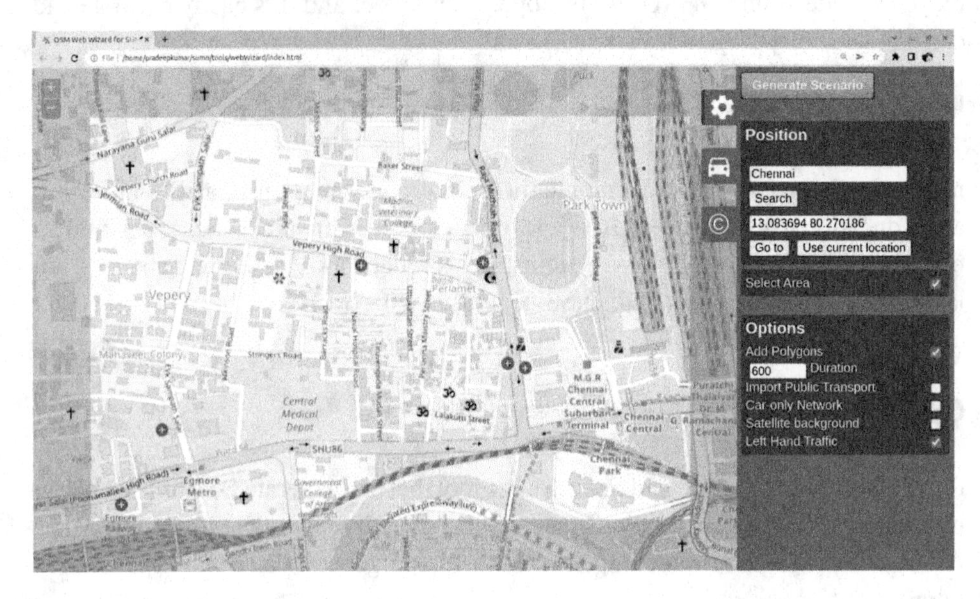


The list of vehicles included in this simulation is cars, motorbikes, buses and pedestrian. There are other types of vehicles also supported, but in this simulation use of these 4 types is selected. Once the “Generate Scenario” button is pressed, a window pops out with a typical graphical UI that shows the traffic pattern as shown in Figure 3. This figure shows the map of a place which is selected through the Lat and Lon as indicated in Figure 2.

*Figure 3. SUMO GUI Output*

From the Figure 3, once the simulation time and parameters are selected, the floating cars (FC) can be seen running on the roads, waiting for the signals, etc. as shown in Figure. 4. The actual image of Figure 4 is zoomed to show the actual cars and buses running on the road.

*Figure 4. Floating Car Data (FCD) in SUMO*

Also while running the simulation, based on the maps, OSM suggested some properties of the road selected. For example, the property might be, "How many dead end junctions" are available in the map, or "how many priority junctions are there in the map" or "how many traffic light junctions are there in the map", etc. This information is shown is Figure 5.

*Figure 5. Summary of the junctions selected in OSM.*

```
Summary:
 Node type statistics:
  Unregulated junctions       : 0
  Dead-end junctions          : 116
  Priority junctions          : 212
  Right-before-left junctions : 76
  Traffic light junctions      : 2
 Network boundaries:
  Original boundary  : 79.84,13.05,80.33,13.56
  Applied offset     : -418519.94,1442448.57
  Converted boundary : 0.00,-6413.93,3018.74,0.00
```

Figure 5 also shows the original boundary selected with Latitude and longitude and the converted boundary for the simulations. This summary helps the researchers to predict the road structures and the maps accordingly before the simulation.

*Table 1. Parameters set for the OSM and SUMO*

| Name of the Parameter | Values |
|---|---|
| Traffic modelling software | SUMO<br>(Simulation of Urban Mobility) |
| Maps | Open Street Maps (OSM) |
| Total number of vehicles | 240+ (generated at random) |
| Total time duration | 600 seconds |
| <Latitude, longitude> | <79.84, 13.56> |
| Type of vehicles | Cars, Buses, motorbikes and pedestrian |
| Type of Traffic | Left hand |

The simulated output from SUMO can be viewed as a floating car data in the screen and to get that data in to a discrete event simulator like ns2 or ns3, the conversion should happen in the following way.

- Conversion of the **.sumocfg** (Sumo Config) file to ns2 mobility file (**mobility. tcl**)
- Ns3 has a facility to include the mobility.tcl file while running its simulations.
- The mobility.tcl file mainly contains the time, mobility of nodes and the locations it moves in a given instant of time is mentioned.
- Once the ns3 script runs, it generates two types of files namely an XML file for NetAnim (Network Animation) and some log files or trace files for plotting the characteristics.

## Converting From SUMO to Network Simulators

As the SUMO software shows only the mobility of the vehicles and properties of the road conditions, it is necessary to convert the libraries to network simulators for further analysis of networks and its performance metrics. Only then, it becomes a vehicular networks. SUMO does the provision of converting to many network simulators like NS2, NS3, OMNeT++, QualNet, etc. with an easier conversion through python scripts. In the network simulators, the parameters like Mac protocols, transport layer protocols, routing protocols, application layer protocols, etc. can be set as per the requirements. Table 2 shows the important parameters that are set for simulating a vehicular network.

*Table 2. Network Parameters for VANETs for NS3*

| Name of the Parameter | Value |
|---|---|
| Mac Protocol | IEEE 802.11p |
| Propagation loss Model | Friss and Two Ray Ground |
| Application Traffic | BSM (Basic Safety Messages) |
| Routing Protocols or Traffic | AODV, DSR, DSDV, OLSR |
| Mobility Model | Random Way Point |
| Scenarios | Two scenarios<br>1. 10 seconds with 50 nodes with a 300X1500m area<br>2. 300 seconds with 100 nodes with a real world map |
| Tracing | ASCII and PCAP |
| Environment | WAVE(Wireless Access in Vehicular Environment) |
| Total Simulated time | 40 seconds or 300 seconds in two scenarios |
| Frequency | 5.9GHz for 802.11p and<br>2.4Ghz for 802.11b |

Various simulated experiments were carried out and the network animation is shown in Figure 6. It shows the vehicles or the nodes in the network through the software NetAnim of NS3. NetAnim reads the xml file and shows the nodes properties and behaviour.

*Figure 6. The nodes in NetAnim of NS3*

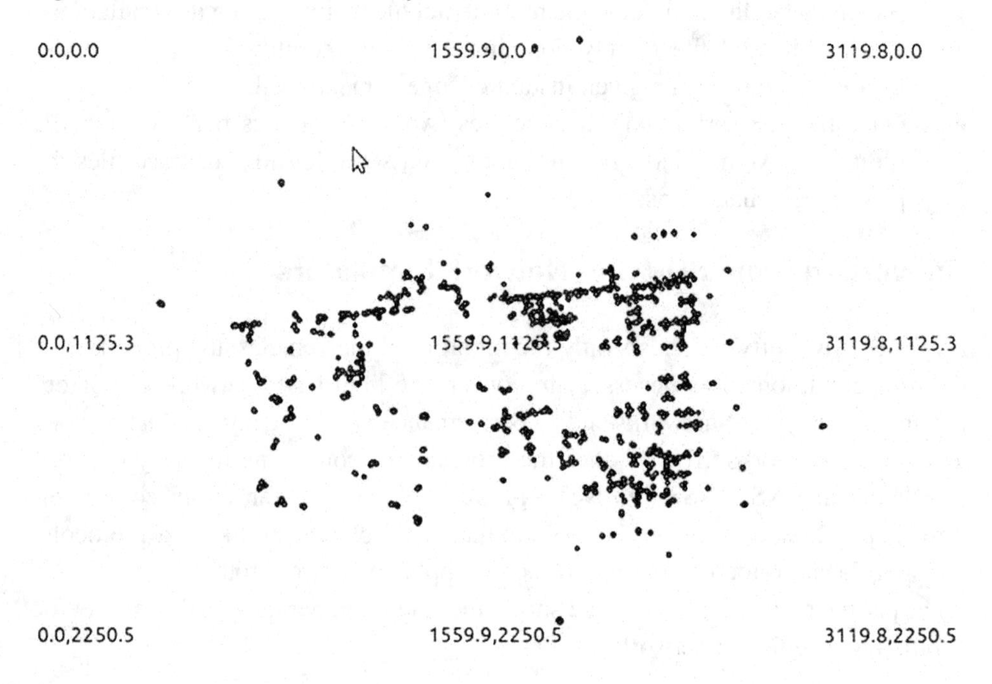


It shows the nodes IP address, MAC address, Transport layer protocol, etc. and the nodes can also move as per the timing specified by the mobility.tcl file. The nodes can also exchange application layer packets between them and can measure throughput, end to end delay, packet delivery ratio, etc. Some of the characteristics have been plotted using the trace files obtained through the simulation. For example, Figure 7 shows the throughput of the network for the 4 protocols and the throughput is good for DSR and then to AODV protocol. Usually AODV protocol outperforms other protocol in many metrics, but in this network the DSR protocol have a slight edge over the AODV protocol. Since the vehicles are moving at high speed, there are some packet losses as well, hence the throughput also get affected.

*Figure 7. Throughput of the network in Kbps*

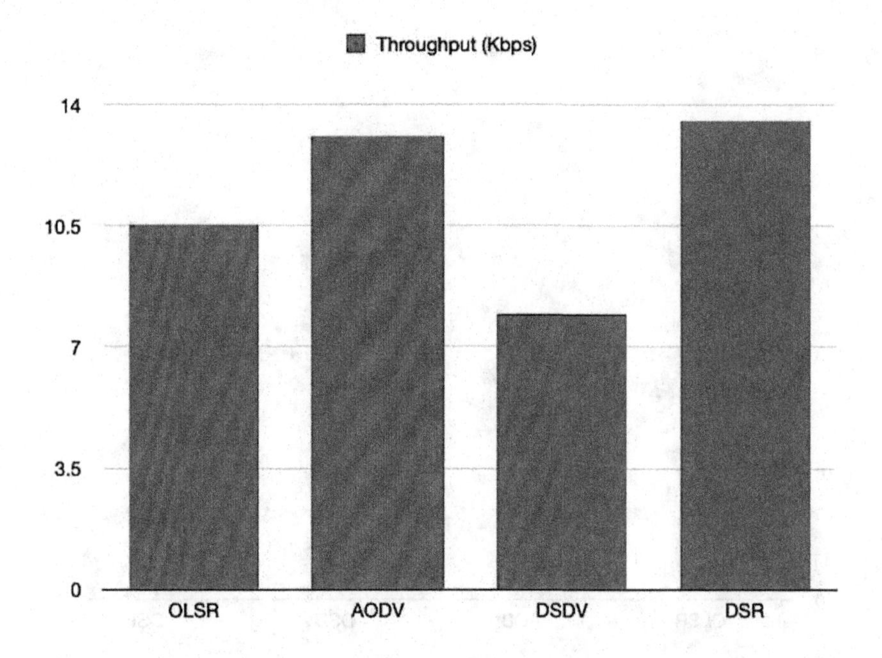


Figure 8 and Figure 9 shows the Mac and physical layer overhead due to these protocols which might affect the radio and communication path between the vehicles. Here also the AODV protocol' overhead is high as there is a higher chance of path establishment as the nodes are moving at high speed. Since AODV select the path dynamically every time, there might be chances are more for the overhead messages as well. But DSR performs well here as well, as the path or the route is known apriori, so there is no need for establishment of the path.

*Figure 8. Mac PHY Overhead (ratio)*

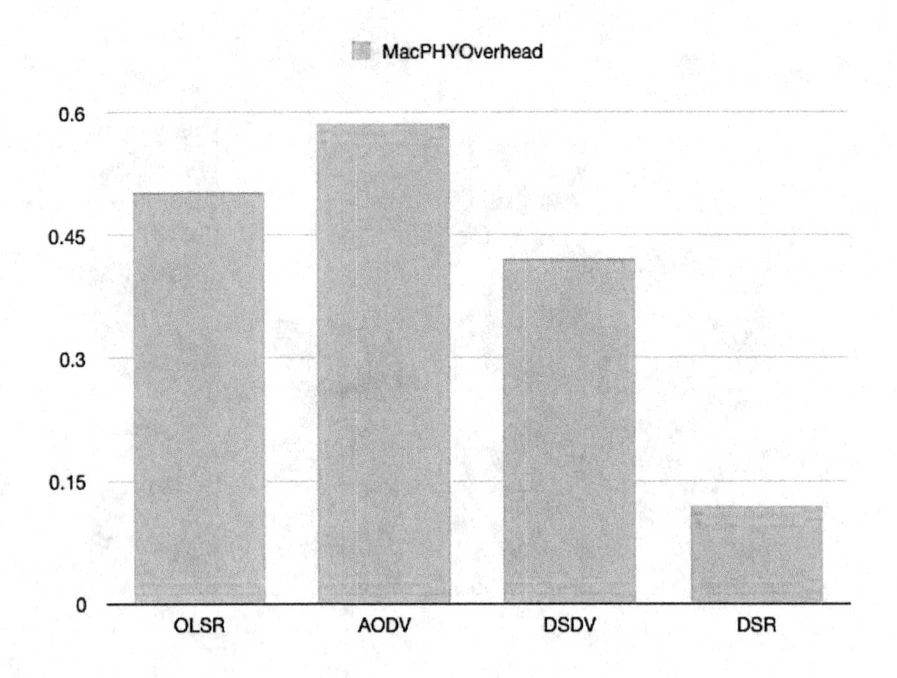


*Figure 9. Mac PHY Overhead*

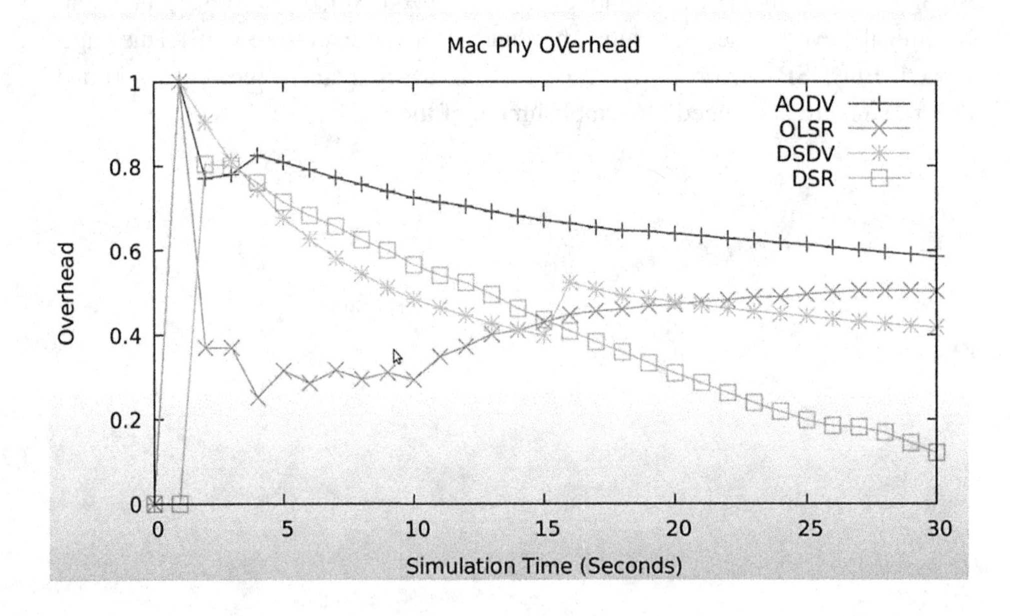

Figure 10 shows the packet receive rate and number of packets received during the traffic generation of nodes. Here also the AODV and DSR protocol performed well as higher the receive rate, better is the network. Also higher receive rate indicates lower packet drop and lower packet loss which indicates it's a better network. Hence the DSR and AODV did well in receive rate.

*Figure 10. Receive Rate comparison between routing protocols*

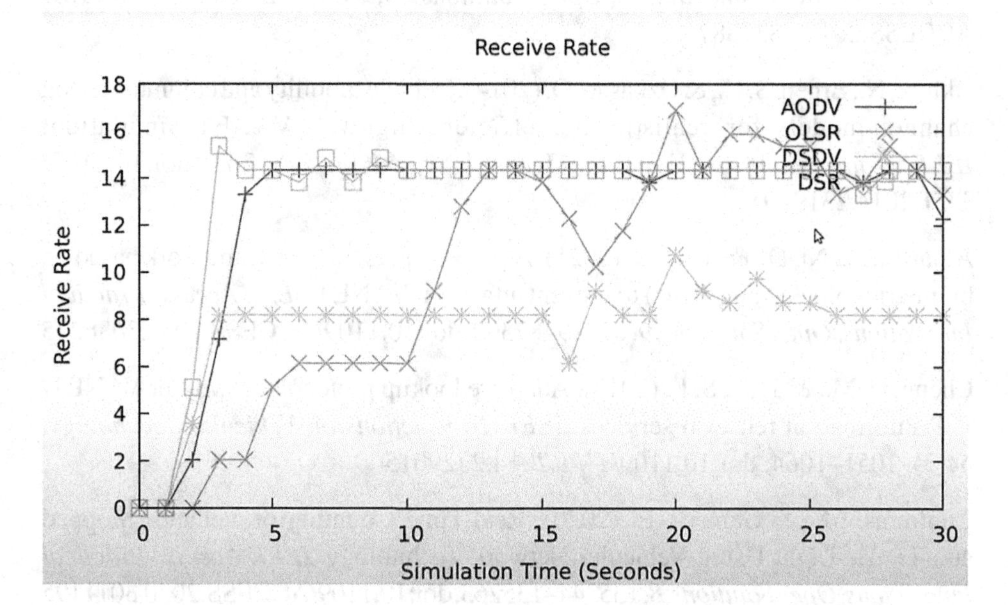


## CONCLUSION

In this chapter, the real-world simulation of SUMO is discussed and how SUMO generates vehicular traffic and how the conversion of vehicles to nodes in a network through the mobility of SUMO is discussed. Various experimental setup were carried out and analytical studies also done. Different performance metrics like throughput, receive rate, mac PHY overhead were plotted against the routing protocols namely AODV, DSR, DSDV and OLSR. Among all these things, the AODV and DSR protocols perform well in many of these metrics. This chapter only tells the importance of sumo and its usage in network modelling.

## REFERENCES

Abuashour, A., & Kadoch, M. (2017). Performance improvement of cluster-based routing protocol in VANET. *IEEE Access: Practical Innovations, Open Solutions*, *5*, 15354–15371. doi:10.1109/ACCESS.2017.2733380

Ahmad, F., Franqueira, V. N., & Adnane, A. (2018). TEAM: A trust evaluation and management framework in context-enabled vehicular ad-hoc networks. *IEEE Access: Practical Innovations, Open Solutions*, *6*, 28643–28660. doi:10.1109/ACCESS.2018.2837887

Akhtar, N., Ergen, S. C., & Ozkasap, O. (2014). Vehicle mobility and communication channel models for realistic and efficient highway VANET simulation. *IEEE Transactions on Vehicular Technology*, *64*(1), 248–262. doi:10.1109/TVT.2014.2319107

Alharthi, A., Ni, Q., & Jiang, R. (2021). A privacy-preservation framework based on biometrics blockchain (BBC) to prevent attacks in VANET. *IEEE Access: Practical Innovations, Open Solutions*, *9*, 87299–87309. doi:10.1109/ACCESS.2021.3086225

Cheng, C. M., & Tsao, S. L. (2014). Adaptive lookup protocol for two-tier VANET/P2P information retrieval services. *IEEE Transactions on Vehicular Technology*, *64*(3), 1051–1064. doi:10.1109/TVT.2014.2329015

Contreras, M., & Gamess, E. (2020). Real-Time Counting of Vehicles Stopped at a Traffic Light Using Vehicular Network Technology. *IEEE Access: Practical Innovations, Open Solutions*, *8*, 135244–135263. doi:10.1109/ACCESS.2020.3011195

Haghighi, M. S., & Aziminejad, Z. (2019). Highly anonymous mobility-tolerant location-based onion routing for VANETs. *IEEE Internet of Things Journal*, *7*(4), 2582–2590. doi:10.1109/JIOT.2019.2948315

Kumar, T. P., & Krishna, P. V. (2021). A survey of energy modeling and efficiency techniques of sensors for IoT systems. *International Journal of Sensors, Wireless Communications and Control*, *11*(3), 271–283. doi:10.2174/221032791099920061 4001521

Kumbhar, F. H., & Shin, S. Y. (2020). DT-VAR: Decision tree predicted compatibility-based vehicular ad-hoc reliable routing. *IEEE Wireless Communications Letters*, *10*(1), 87–91. doi:10.1109/LWC.2020.3021430

Liu, B., Jia, D., Wang, J., Lu, K., & Wu, L. (2015). Cloud-assisted safety message dissemination in VANET–cellular heterogeneous wireless network. *IEEE Systems Journal, 11*(1), 128-139.

Luo, G., Li, J., Zhang, L., Yuan, Q., Liu, Z., & Yang, F. (2018). sdnMAC: A software-defined network inspired MAC protocol for cooperative safety in VANETs. *IEEE Transactions on Intelligent Transportation Systems*, *19*(6), 2011–2024. doi:10.1109/TITS.2017.2736887

Seliem, H., Shahidi, R., Ahmed, M. H., & Shehata, M. S. (2018). Drone-based highway-VANET and DAS service. *IEEE Access: Practical Innovations, Open Solutions*, *6*, 20125–20137. doi:10.1109/ACCESS.2018.2824839

Shah, A. S., Karabulut, M. A., Ilhan, H., & Tureli, U. (2020). Performance optimization of cluster-based MAC protocol for VANETs. *IEEE Access: Practical Innovations, Open Solutions*, *8*, 167731–167738. doi:10.1109/ACCESS.2020.3023642

Tiennoy, S., & Saivichit, C. (2018). Using a distributed roadside unit for the data dissemination protocol in VANET with the named data architecture. *IEEE Access: Practical Innovations, Open Solutions*, *6*, 32612–32623. doi:10.1109/ACCESS.2018.2840088

Wang, J., Chen, H., & Sun, Z. (2020). Context-Aware Quantification for VANET Security: A Markov Chain-Based Scheme. *IEEE Access: Practical Innovations, Open Solutions*, *8*, 173618–173626. doi:10.1109/ACCESS.2020.3017557

# Chapter 7
# Review on Recent Applications of Internet of Flying Things

**Vanitha Veerasamy**
*Kumaraguru College of Technology, India*

**Rajathi Natarajan**
*Kumaraguru College of Technology, India*

## ABSTRACT

*Unmanned aerial vehicles (UAVs), typically known as drones, are aerial machines that can be programmed and controlled remotely using mobile devices and are connected via wireless communication technology. Because of their ease of deployment, dynamic configuration, low maintenance costs, high mobility, and faster reaction, they are becoming more widely used in a variety of applications. As a result, a new paradigm known as flying ad hoc networks (FANETs) has emerged, which is a subset of mobile ad hoc networks with special aviation-related properties. FANET ideas have been combined with the internet of things (IoT), resulting in the internet of flying things (IoFT), a paradigm that enables a significant new level of applications, solves existing challenges in UAVs and IoT, and broadens the spectrum of potential uses. This study focuses on various IoFT applications and challenges in IoFT implementation.*

## INTRODUCTION

The Internet of Things (IoT) is a prominent technology with applications in different domains. Millions of devices are connected to each other and to the Internet in an IoT system. Healthcare monitoring, remote patient monitoring, environmental

DOI: 10.4018/978-1-6684-3610-3.ch007

monitoring, precision agriculture, energy monitoring and transportation systems are all examples of where IoT is applied. In addition, the Internet of Things (IoT) is a key technology for creating smart cities.

Now-a-days Unmanned Aerial Vehicles (UAVs) are used for variety of military and civilian applications. UAVs are small size aircrafts. A UAV is made up of a variety of sensors, actuators, compute units and storage units. The combination of UAVs and the IoTs is a new emerging direction for academia and industry. IoT allows devices to connect to any network, at any time, to provide any service. UAVs can be used for a variety of functions in UAV-based IoT. UAV based IoT is called as Internet of Flying Things.

This paper is organized as follows: Section 2 delves detailed introduction to Internet of Flying Things (IoFT), section 3 deals with IoFT applications. The section 4 lists the challenges and chapter 5 gives the concluding remarks.

## OVERVIEW OF INTERNET OF FLYING THINGS

The IoFT is an emerging area having wide variety of application possibilities. On the Internet of Things, the UAV is a critical component. The following section explains about UAV, multiple-UAV systems, FANETs, and IoFT in detail.

### Unmanned Aerial Vehicle (UAV)

The rapid technological advancement in the communication technologies, avionics and micro electromechanical system has paved the path to new UAV systems. UAV is also known as Drone, is a flying vehicle or an aircraft. UAVs can fly autonomously with the help of on-board computer and usually without human or it can also be operated remotely by a human operator (Hassanalian & Abdelkefi, 2017). These UAVs are capable of flying a few thousands of kilometres and also capable of carrying lethal or nonlethal payloads. The UAVs are categorized into different types based on weight and range, landing, rotors etc. The classification of UAVs is shown in the figure 1.

The characteristics of UAVs include easy installation and relatively small operating expenses. The UAVs are used in a variety of situations where the presence of humans is difficult, impossible, or dangerous. They are capable for performing in both outdoor and indoor locations as well as in very challenging environments. UAVs are used in both military and civilian applications. In military, these UAVs provide an insight about the specific areas easily. UAVs mounted with specialized cameras are capable of providing quality images even about the dark areas. Small size micro UAVs are mainly used to investigate indoor areas. Micro UAVs are used to perform investigations inside the buildings. UAVs are also used for search and rescue operations, disaster monitoring, remote sensing, managing wildfire, base station for communication and traffic monitoring. They are also capable of performing missions in oceans or in other planets.

*Figure 1. Classification of UAV*

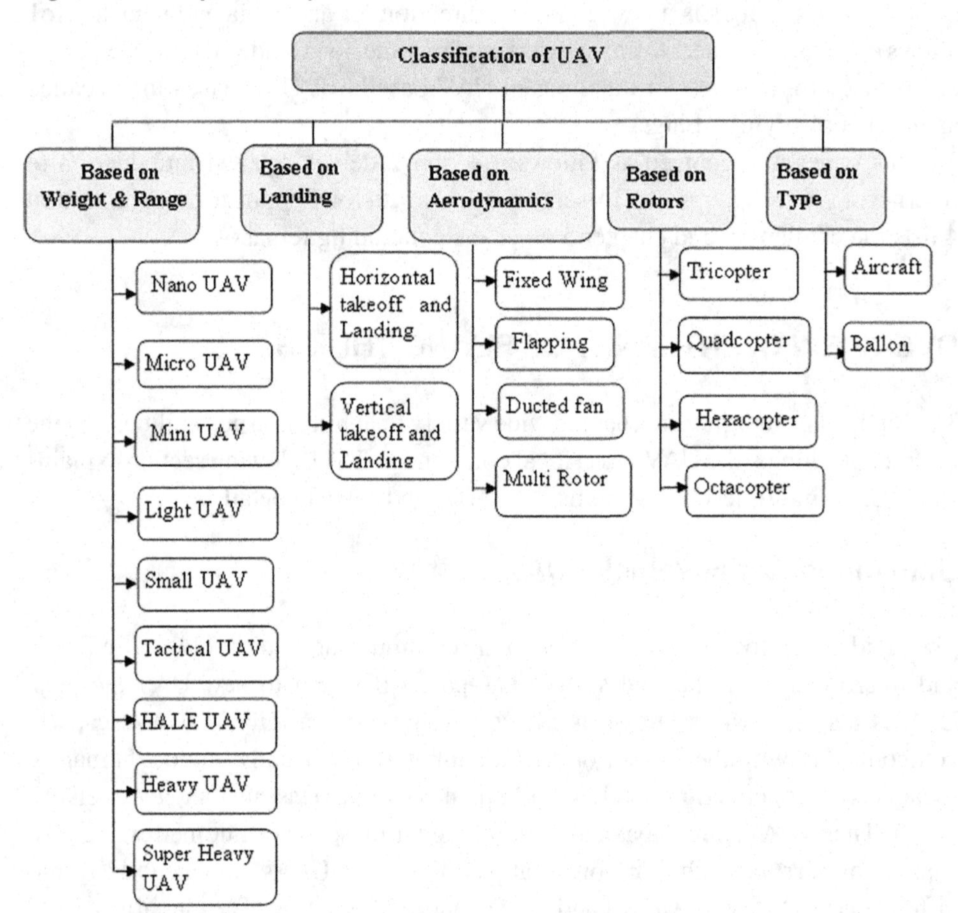


## Multi-UAVs

Single-UAV system has some performance limitation. Instead of using single highly configured UAV, networking multiple simple UAVs has numerous advantages compared to single-UAV (Skorobogatov et al., 2020). Some of the advantages of multi-UAV system are:

- Expenditure: Small size UAVs are cheaper compared larger one.
- Scalability: Coverage of multi-UAV system is higher than single-UAV systems.
- Survivability: Multi-UAV system continues to operate even if one or two UAV doesn't work.
- Speed: Collection of UAVs completes the task very quickly.

However, multi-UAV communication system has some drawbacks like: (i) UAVs need to be equipped with specialized devices to communicate with a ground base station or to a satellite system. (ii) UAVs may have intermittent communication link. (iii) UAVs have range restriction to communicate with the ground base station. To address these problems Flying Adhoc Network was proposed.

## Flying Adhoc Networks

Alternative network architecture for multi-UAV systems is to create an ad hoc network. This special kind of ad hoc network architecture is called as Flying Ad Hoc Networks (FANETs) (Bekmezci et al., 2013). Flying Adhoc Networks provides reliable communication link. It is capable of communicating with ground base quickly without any difficulty (Kumari et al., 2015).

FANET is a type of MANET and has number of common characteristics. But there are also some differences between FANET and ad hoc networks:

- In FANET, nodes mobility is much higher compared to other adhoc networks. Due to this the topology also changes more frequently. Hence maintaining connectivity becomes an important issue for FANETs.
- For UAV coordination and collaboration, FANET leverages peer-to-peer connections. It collects data about various physical phenomena and passes to the ground control station. Hence, peer-to-peer communication as well as communication with ground station both are possible.
- Sparse FANET nodes cannot communicate long distance. High quality antenna is required to transmit data for long distance.
- FANET requires very high bandwidth to transmit real time data.
- Multi-UAV systems may produce different types of data
  FANET has several challenges as well (Singh, 2015):
- In FANET node density is low and mobility of the UAVs is high. This leads to decrease in the network performance.
- FANET nodes depend on the battery power. Hence, the selecting a suitable UAV challenging task.
- Transmitting the data to the ground station is an important FANET issue.

## Internet of Flying Things

IoT is a network of Internet connected devices that send embedded sensor data to the cloud for processing. Currently, IoT faces the some issues: (i) Replacing the battery of the IoT devices is the main challenge. (ii) IoT coverage is limited especially under adverse weather conditions. (iii) IoT devices have limited capability (iv) IoT devices depends on battery power. Hence, managing such constrained devices is a key issue needs to be addressed.

UAVs are usually planned for specific applications and service. In addition to their original tasks, UAVs can also be used for number of other services specifically in the IoT domain. In this way, UAVs form an innovative UAV-based IoT platform operational in the sky (Zaidi et al., 2021). This will reduce the costs for developing a novel ecosystem.

IoFT can collect data using sensors available in the UAVs. When mounted with suitable IoT devices, cameras and transceivers UAVs can be used to collect IoT data. The sensors, cameras, actuators, and RFIDs placed in the UAVs are remotely controllable. IoT data can be processed in the UAV itself or transmitted to a server (Motlagh et al., 2017; Motlagh et al., 2016).

Figure 2 shows the high-level architecture of IoFT system. UAVs may form clusters and elect cluster heads. In a cluster, one of the UAV may function as a cluster head and transfer the collected IoT data to the ground control station.

IoFT architecture consists of system orchestrator to coordinate UAVs. It keeps track of all information about UAVs. System orchestrator is an intelligent device, responsible for ensuring security.

*Figure 2. IoFT Architecture*

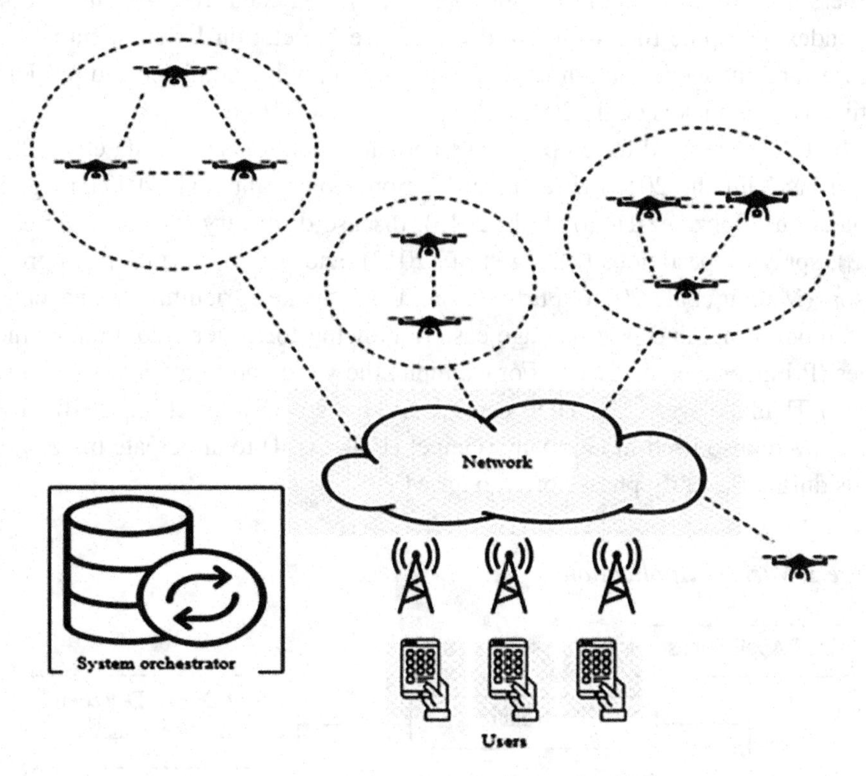


## APPLICATIONS

IoFT have been successfully used in various domains including agriculture, healthcare, traffic monitoring, pollution monitoring, disaster monitoring, search and rescue operations. This section summarises some of the most recent research on IoFT applications. Figure 3 show the various application domains of IoFT.

### Agriculture

Agriculture is one of the most common areas where IoFT can be used to improve crop yield while reducing environmental impact. The world's population is predicted to reach 9.8 billion people in 2050, a 25% increase over the current amount (UN, 2017). It is vital to supply the ever-increasing population's food needs. Farmers must visit agriculture locations on a regular basis during the crop life to have a better sense of the crop conditions in traditional farming. As a result, smart agriculture is essential, because the majority of farm time is spent in monitoring and assessing crop status rather than performing actual field work (Navulur & Prasad, 2017).

Farmers use drones to count the number of fruits on each tree, measure the leaf area index, compute the Normalized Difference Vegetation Index, identify plant diseases and mange irrigation so that farmers can apply chemicals and pesticides on time (Kakamoukas et al., 2020).

The UAVs are used for crop management and monitoring (Huang et al., 2013; Muchiri & Kimathi, 2016), weed identification (Louargant et al., 2017), irrigation preparation (Gonzalez-Dugo et al., 2013), disease discovery (Garcia-Ruiz et al., 2013), spraying pesticides (Zhang et al., 2017), and data collection from ground sensors (Mathur et al., 2016). Furthermore, UAVs make agricultural management, weed monitoring, and pest damage easier, allowing for faster resolution of these issues (Primicerio et al., 2012). For example, the yield and total biomass of a rice crop in Thailand were estimated using UAV images (Swain et al., 2010). UAV images were also used in Germany (Geipel et al., 2014) to anticipate maize grain yields during the early phases of crop growth.

*Figure 3. Various Applications*

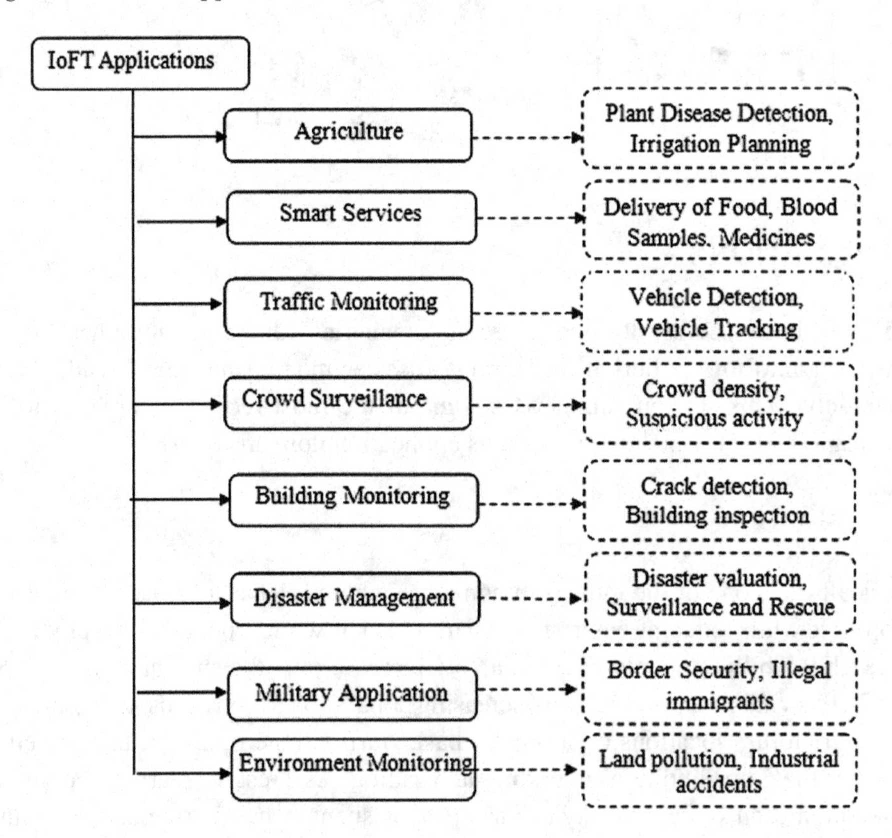

## Smart Services

Food, parcels, and other products can be transported by UAVs (Fandetti, 2017) (PwC, n.d.). Ambulance drones can transport medicines, vaccines and blood samples into and out of hard-to-reach locations in the health-care industry. They can transfer medical instruments quickly in the critical minutes following cardiac arrest. They can also include live video streaming systems, which enable paramedics to remotely watch and instruct on-scene personnel on how to use medical gear (Redstagfulfillment, n.d.a). UAV testing has been conducted by numerous postal organisations to assess the practicality and portability of UAV delivery services (Redstagfulfillment, n.d.b). Quad rotor drones can be used to transport packages. In (Wikipedia, 2017), the authors discussed about commercial drone delivery services.

## Traffic Monitoring

UAVs have been touted as a cutting-edge traffic monitoring system for gathering data on road conditions. UAVs are more cost-effective than typical monitoring systems like security video cameras, microwave sensors, and can monitor huge continuous road segments or a specific road segment (Unmanned Cargo, n.d.). In (Ke et al., 2016), the authors present vehicle detection and tracking system based on imaging data gathered by a UAV. This technology generates the vehicle's dynamic information, such as locations and velocities over time. Drone-based FANETs offer a number of benefits in terms of remote traffic monitoring, including smart surveillance and security (Wang et al., 2016).

The research in (Reshma et al., 2016) focuses on the security challenges associated with UAV-based road traffic monitoring systems. The role of an unmanned aerial vehicle in a road traffic management system is examined in this paper. Instead of using image processing to depict the road network, the authors used sensor networks and graph theory. UAVs were utilised as an alternate data gathering approach by Sutheerakul et al. (Sutheerakul et al., 2017) to monitor pedestrian traffic.

## Crowd Surveillance

Drone usage in crowded circumstances is becoming increasingly fascinating in detecting early indicators of a stampede and congestion. Drone-captured real-time videos and images will cover a larger region and have a customizable alleviation view. UAV with suitable communication modules will be helpful in real-time imagery feed for mission-critical applications.

Motlagh et al. (Motlagh et al., 2017) discussed the use of unmanned aerial vehicles (UAVs) and IoT in crowd surveillance. They demonstrated an IoT-based infrastructure and created a crowd monitoring testbed. The UAV can be used to offer crowd safety, detect criminal activity, employ facial recognition, and many more uses in crowd surveillance. The major goal was to determine how much energy UAVs consumed during onboard processing and ground station processing.

Almagbile et al (Al-Sheary & Almagbile, 2017) proposed and tested a pedestrian crowd monitoring system. They used real-time images captured by UAVs to implement a crowd monitoring system. Image segmentation algorithms were used to calculate crowd density from captured images. In (Almagbile, 2019) the authors used corner detection and cluster analysis to estimate the crowd density from the images collected by UAVs. Chriki et al. (Chriki et al., 2019) used multiple UAVs to monitor a congested location and provided centralised data-oriented communication architecture for crowd monitoring. In (Wang et al., 2013), Wang et al. discussed the use of UAVs and unmanned ground vehicles (UVGs) to monitor intelligence gathering, monitor borders, and control crowds.

## Building Monitoring

UAVs can be utilised for real-time surveillance of construction project sites for construction and infrastructure inspection (Gheisari et al., 2014). As a result, project managers can employ unmanned aerial vehicles to monitor the construction site and acquire a better understanding of its progress without having to personally visit the site. UAV is used to ensure construction site safety by providing real-time visual images of the site. Crack detection and surface degradation assessment using a UAV with image processing algorithms (Sankarasrinivasan et al., 2015). In North America, Industrial SkyWorks (Industrial Skyworks, n.d.) employs drones for building inspections and oil/gas inspections.

## Disaster Management

Because of the size of the UAVs, they can reach regions that would be difficult for humans to reach in the event of a crisis. During a disaster event, UAVs could provide rescuers with a bird's eye view of the disaster area, which is very important for a comprehensive disaster management system that involves data collection, victim localization and rescue optimization. UAVs can be used in conjunction with existing early warning systems in the pre-disaster stage to precisely estimate the outbreak timing and scope, reducing the economic and material costs that a disaster may cause through a disaster prevention strategy.

UAVs can give high-resolution real-time images of even inaccessible areas during a disaster, which can be utilised to develop precise hazard maps to direct the rescuers to assess the situation, planning relief, and conducting rescue. After disaster,

UAVs can be used to map the affected areas in high resolution within a short time for efficient responses. The authors of (Rajan et al., 2021) discussed the usage of UAVs for disaster evaluation. In (Asadpour et al., 2013), the authors presented the challenges encountered in search and rescue missions.

### Military Services

Military services benefit greatly from IoFT. In the military, setting up a proper communication system is quite tough. Flying devices can be used to spot border crossings and follow illegal immigrants due to advancements in automated monitoring. The military personnel primarily use for communication between soldiers or between their barracks. Threats, jamming, spoofing, and other attacks are all possible on the flying devices. It is necessary to ensure the security protocol for communication between UAVs and between a UAV and a Ground Control Station to be more secured.

### Environmental Monitoring

Because of their capacity to enter hazardous regions, UAVs can be used to examine atmospheric composition, air quality, and climate characteristics. Alvear et al. (Alvear et al., 2017) proposed the use of UAVs for pollution monitoring. The flying things can be utilised for wildlife monitoring, forest fire detection, land pollution monitoring as well as in oil and water pipelines.

## CHALLENGES IN IOFT IMPLEMENTATION

Though the IoFTs are used in many application areas, they face a number of challenges due to their restricted energy and processing capabilities. Figure 4 depicts some of the key challenges faced by the researcher in IoFT implementation.

*Figure 4. Challenges in IoFT implementation*

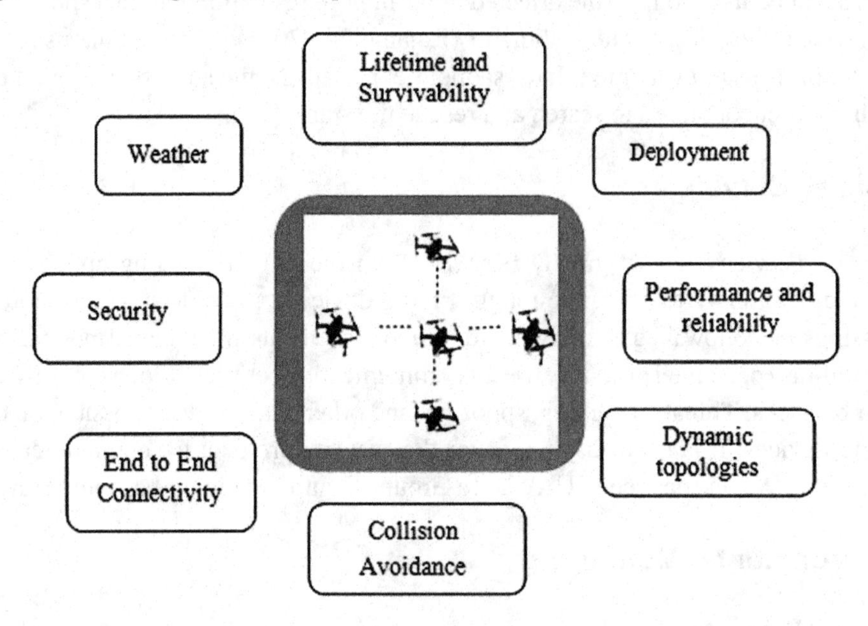


- **Lifetime and Survivability:** The UAVs are battery operated and have a limited power. In addition to communication, the UAV takes energy for computational activities. Improving the survivability and enhancing the lifetime with the limited resources is a major research issue. To increase battery capacity, suitable recharge scheduling and energy harvesting methods needs to be employed.
- **Weather:** UAVs may vary from their intended trajectories due to weather conditions. In the event of a disaster, weather conditions may cause UAVs to fail in their missions. The deployment of UAVs is complicated by bad weather conditions such as high wind, rain, and storm.
- **Security:** Each UAV in IoFT is required to use wireless links to exchange messages and routing information. Therefore, assaults on IoFT are possible. As a result secured routing protocols must be designed and implemented. Eavesdropping and session hijacking are two examples of attacks that could be used on the communications channels between different entities of UAV systems. Direct attacks on UAVs can inflict considerable system damage.
- **End-to-end connectivity:** The connection between the source and the destination should always be maintained in the IoFT system. For improved coverage and cooperation, some application areas, such as real-time monitoring, require substantial end-to-end connection.

- **Collision Avoidance:** Collision avoidance is one of the issue that UAVs face. An UAV must be able to avoid collisions with both static and moving obstacles. Collisions can occur for a variety of causes, including operator error, equipment failure, or adverse weather conditions. One of the most difficult problems for autonomous vehicles is detecting and avoiding impediments, which is considerably more difficult in dynamic situations with several UAVs and moving obstacles. Once a UAV has received information about the barriers, it must decide whether any collisions are imminent. The necessity for intelligent and highly dependable collision avoidance systems is evident and incontrovertible from the standpoint of public safety, given the rising use of unmanned vehicles and, in particular, the exponential development in the uses of UAVs in public areas and our daily lives.
- **Dynamic topologies:** One of the most significant difficulties is highly dynamic topology of UAV networks, which regularly disconnects the link. As a result of the high mobility and dynamic topology of UAVs, developing a routing system that ensures reliable communication in UAV networks is difficult. For UAV coordination in FANET, a suitable localization approach for estimating the position of high-speed UAV nodes, as well as adaptive routing methods, should be considered.
- **Performance and reliability:** The performance of IoFT is accomplished with the mission completion time and efficient resource utilization. To understand resource efficiency and reliability, consumption-focused indicators should be incorporated. The monitoring of consumption focused resources is evaluated in terms of performance. The efficient use of resources leads to successful missions. Therefore, the resource utilization is considered as a significant parameter for performance and reliability in IoFT.
- **Deployment Strategies:** The position of UAVs should be based on some environmental condition in many cases. If we're tracking wild creatures in the jungle, the UAVs should be positioned to keep an eye on as many as feasible. It is critical to have an efficient UAV deployment framework with the goal of minimizing the number of UAVs and transmission power required to meet the users' rate requirements.

## CONCLUSION

IoFT is becoming a potential field to efficiently support real-time IoT applications by integrating UAVs with IoT. This study presents a comprehensive review on IoFT applications and implementation challenges. Furthermore, more research is to be carried out to investigate the new IoFT scheme that overcomes the IoFT issues to provide reliable and efficient computation as well as the transmission and storage of UAV-collected data to Internet servers.

## REFERENCES


Al-Sheary, A., & Almagbile, A. (2017). Crowd monitoring system using unmanned aerial vehicle (UAV). *Journal of Civil Engineering and Architecture*, *11*(11), 1014–1024. doi:10.17265/1934-7359/2017.11.004

Almagbile, A. (2019). Estimation of crowd density from UAVs images based on corner detection procedures and clustering analysis. *Geo-Spatial Information Science*, *22*(1), 23–34. doi:10.1080/10095020.2018.1539553

Alvear, O., Zema, N. R., Natalizio, E., & Calafate, C. T. (2017). Using UAV-based systems to monitor air pollution in areas with poor accessibility. *Journal of Advanced Transportation*, *2017*. doi:10.1155/2017/8204353

Asadpour, M., Giustiniano, D., Hummel, K. A., & Heimlicher, S. (2013, July). Characterizing 802.11 n aerial communication. In *Proceedings of the second ACM MobiHoc workshop on Airborne networks and communications* (pp. 7-12). 10.1145/2491260.2491262

Bekmezci, I., Sahingoz, O. K., & Temel, Ş. (2013). Flying ad-hoc networks (FANETs): A survey. *Ad Hoc Networks*, *11*(3), 1254–1270. doi:10.1016/j.adhoc.2012.12.004

Chriki, A., Touati, H., Snoussi, H., & Kamoun, F. (2019, June). UAV-GCS centralized data-oriented communication architecture for crowd surveillance applications. In *2019 15th International Wireless Communications & Mobile Computing Conference (IWCMC)* (pp. 2064-2069). IEEE.

Fandetti, G. N. (2017). *Method of drone delivery using aircraft*. U.S. Patent Application 14/817,356.

Garcia-Ruiz, F., Sankaran, S., Maja, J. M., Lee, W. S., Rasmussen, J., & Ehsani, R. (2013). Comparison of two aerial imaging platforms for identification of Huanglongbing-infected citrus trees. *Computers and Electronics in Agriculture*, *91*, 106–115. doi:10.1016/j.compag.2012.12.002

Geipel, J., Link, J., & Claupein, W. (2014). Combined spectral and spatial modeling of corn yield based on aerial images and crop surface models acquired with an unmanned aircraft system. *Remote Sensing*, *6*(11), 10335–10355. doi:10.3390/rs61110335

Gheisari, M., Irizarry, J., & Walker, B. N. (2014). UAS4SAFETY: The potential of unmanned aerial systems for construction safety applications. In *Construction Research Congress 2014: Construction in a Global Network* (pp. 1801-1810). Academic Press.

Gonzalez-Dugo, V., Zarco-Tejada, P., Nicolás, E., Nortes, P. A., Alarcón, J. J., Intrigliolo, D. S., & Fereres, E. J. P. A. (2013). Using high resolution UAV thermal imagery to assess the variability in the water status of five fruit tree species within a commercial orchard. *Precision Agriculture*, *14*(6), 660–678. doi:10.100711119-013-9322-9

Hassanalian, M., & Abdelkefi, A. (2017). Classifications, applications, and design challenges of drones: A review. *Progress in Aerospace Sciences*, *91*, 99–131. doi:10.1016/j.paerosci.2017.04.003

Huang, Y., Thomson, S. J., Hoffmann, W. C., Lan, Y., & Fritz, B. K. (2013). Development and prospect of unmanned aerial vehicle technologies for agricultural production management. *International Journal of Agricultural and Biological Engineering*, *6*(3), 1–10.

Industrial Skyworks. (n.d.). *Drone inspection services*. https://industrialskyworks.com/droneinspections-services

Kakamoukas, G. A., Sarigiannidis, P. G., & Economides, A. A. (2020). FANETs in Agriculture-A routing protocol survey. *Internet of Things*, 100183.

Kazmi, W., Bisgaard, M., Garcia-Ruiz, F., Hansen, K. D., & la Cour-Harbo, A. (2011). Adaptive surveying and early treatment of crops with a team of autonomous vehicles. In *Proceedings of the 5th European Conference on Mobile Robots ECMR 2011* (pp. 253-258). Academic Press.

Ke, R., Li, Z., Kim, S., Ash, J., Cui, Z., & Wang, Y. (2016). Real-time bidirectional traffic flow parameter estimation from aerial videos. *IEEE Transactions on Intelligent Transportation Systems*, *18*(4), 890–901. doi:10.1109/TITS.2016.2595526

Kumari, K., Sah, B., & Maakar, S. (2015). A survey: Different mobility model for FANET. *International Journal of Advanced Research in Computer Science and Software Engineering*.

Louargant, M., Villette, S., Jones, G., Vigneau, N., Paoli, J. N., & Gée, C. (2017). Weed detection by UAV: Simulation of the impact of spectral mixing in multispectral images. *Precision Agriculture*, *18*(6), 932–951. doi:10.100711119-017-9528-3

Mathur, P., Nielsen, R. H., Prasad, N. R., & Prasad, R. (2016). Data collection using miniature aerial vehicles in wireless sensor networks. *IET Wireless Sensor Systems*, *6*(1), 17–25. doi:10.1049/iet-wss.2014.0120

Motlagh, N. H., Bagaa, M., & Taleb, T. (2017). UAV-based IoT platform: A crowd surveillance use case. *IEEE Communications Magazine*, *55*(2), 128–134. doi:10.1109/MCOM.2017.1600587CM

Motlagh, N. H., Taleb, T., & Arouk, O. (2016). Low-altitude unmanned aerial vehicles-based internet of things services: Comprehensive survey and future perspectives. *IEEE Internet of Things Journal*, *3*(6), 899–922. doi:10.1109/JIOT.2016.2612119

Muchiri, N., & Kimathi, S. (2016, June). A review of applications and potential applications of UAV. In *Proceedings of sustainable research and innovation conference* (pp. 280-283). Academic Press.

Navulur, S., & Prasad, M. G. (2017). Agricultural management through wireless sensors and internet of things. *Iranian Journal of Electrical and Computer Engineering*, *7*(6), 3492. doi:10.11591/ijece.v7i6.pp3492-3499

Primicerio, J., Di Gennaro, S. F., Fiorillo, E., Genesio, L., Lugato, E., Matese, A., & Vaccari, F. P. (2012). A flexible unmanned aerial vehicle for precision agriculture. *Precision Agriculture*, *13*(4), 517–523. doi:10.100711119-012-9257-6

PwC. (n.d.). *Emerging technology*. http://usblogs.pwc.com/emergingtechnology/

Rajan, J., Shriwastav, S., Kashyap, A., Ratnoo, A., & Ghose, D. (2021). Disaster management using unmanned aerial vehicles. In *Unmanned Aerial Systems* (pp. 129–155). Academic Press. doi:10.1016/B978-0-12-820276-0.00013-3

Redstagfulfillment. (n.d.a). *The future of distribution*. https://redstagful_llment.com/the-future-ofdistribution/

Redstagfulfillment. (n.d.b). https://redstagful_llment.com

Reshma, R., Ramesh, T., & Sathishkumar, P. (2016, January). Security situational aware intelligent road traffic monitoring using UAVs. In *2016 international conference on VLSI systems, architectures, technology and applications (VLSI-SATA)* (pp. 1-6). IEEE.

Sankarasrinivasan, S., Balasubramanian, E., Karthik, K., Chandrasekar, U., & Gupta, R. (2015). Health monitoring of civil structures with integrated UAV and image processing system. *Procedia Computer Science*, *54*, 508–515. doi:10.1016/j.procs.2015.06.058

Singh, S. K. (2015). A comprehensive survey on FANET: Challenges and advancements. *International Journal of Computer Science and Information Technologies*, *6*(3), 2010–2013.

Skorobogatov, G., Barrado, C., & Salamí, E. (2020). Multiple UAV systems: A survey. *Unmanned Systems*, *8*(02), 149–169. doi:10.1142/S2301385020500090

Sutheerakul, C., Kronprasert, N., Kaewmoracharoen, M., & Pichayapan, P. (2017). Application of unmanned aerial vehicles to pedestrian traffic monitoring and management for shopping streets. *Transportation Research Procedia*, *25*, 1717–1734. doi:10.1016/j.trpro.2017.05.131

Swain, K. C., Thomson, S. J., & Jayasuriya, H. P. (2010). Adoption of an unmanned helicopter for low-altitude remote sensing to estimate yield and total biomass of a rice crop. *Transactions of the ASABE*, *53*(1), 21–27. doi:10.13031/2013.29493

UN. (2017). *World population prospects*. https://www.un.org/development/desa/en/news/population/world-population-prospects-2017.html

Unmanned Cargo. (n.d.). *Drones going postal*. http://unmannedcargo.org/drones-going-postal-summary-postal-servicedelivery- drone-trials/

Wang, L., Chen, F., & Yin, H. (2016). Detecting and tracking vehicles in traffic by unmanned aerial vehicles. *Automation in Construction*, *72*, 294–308. doi:10.1016/j.autcon.2016.05.008

Wang, Z., Li, M., Khaleghi, A. M., Xu, D., Lobos, A., Vo, C., Lien, J.-M., Liu, J., & Son, Y. J. (2013). DDDAMS-based crowd control via UAVs and UGVs. *Procedia Computer Science*, *18*, 2028–2035. doi:10.1016/j.procs.2013.05.372

Wikipedia. (2017). Delivery Drone. Accessed: Dec. 2017. Available: https://en.wikipedia.org/wiki/Delivery_drone

Zaidi, S., Atiquzzaman, M., & Calafate, C. T. (2021). Internet of flying things (IoFT): A survey. *Computer Communications*, *165*, 53–74. doi:10.1016/j.comcom.2020.10.023

Zhang, D., Jiang, N., Li, H., & Wu, C. (2017, December). IOP Conference Series: Materials Science and Engineering. In *International Conference on Robotics andMechantronics (ICRoM2017)* (*Vol. 12*, p. 14). Academic Press.

# Chapter 8
# A Fusion of VANETs and IoT for Intelligent and Secure Communication

**Ashwani Kant Shukla**
*Babasaheb Bhimrao Ambedkar University, India*

**Vivek Shukla**
*Babasaheb Bhimrao Ambedkar University, India*

**Raj Shree**
*Babasaheb Bhimrao Ambedkar University, India*

**Ravi Prakash Pandey**
*Dr. Rammanohar Lohia Avadh University, India*

**Dhirendra Pandey**
*Babasaheb Bhimrao Ambedkar University, India*

## ABSTRACT

*The modern smart cities completely rely on the technology where the smart and intelligent-based transportation management system is a primary requirement and that can only be achieved by the advancement in traditional vehicular ad-hoc network (VANET). The primary two techniques such as IoT and security mechanism are incorporated with the VANET system which helps to design the robust framework. Due to quick transformation in technological landscape, the threats and attacks also get advanced. The most prominent characteristics of the VANET is to provide the self-aware system which assists for the better management in the transportation system. As per the advancement in the attacks, the robust defense mechanism should also be increased. If not, there would be the huge loss in terms of the lives of living beings, societal, and economic. Therefore, this study is completely based on the analysis and recommendation for developing the robust IOV system, which ensures the secure infrastructure.*

DOI: 10.4018/978-1-6684-3610-3.ch008

## INTRODUCTION

Vehicular Ad hoc network (VANET) deals with the network which is designed using ad-hoc system. For sharing the significant information from one device to another device the moving vehicles and devices have to connect in the common platform over secure wireless network. With vehicles and other equipment created a small network at the same time behaves as nodes in the network whereas if one node keep any of the information it shares to all connected nodes. These processes of sharing the information within the network it continues. In addition, the data received by nodes from different sources serves to refine the information and share it with other connected devices (M. Feiri et. al, 2013). The concept of the sharing and communicating among the connected nodes is based on open network such as the nodes are free to connect and leave the network anytime. Nowadays, the newly developed smart and intelligent based vehicle see running on road is basically equipped with on board sensors which makes quite effortless to connect and integrates the network using the characteristics of the VANET. It is suppose to develop different types of wireless technologies such as dedicated short range communications and others are cellular and satellite based wireless technologies. Also, ad-hoc networks of vehicles can be accounted as a parameter of intelligent transportation frameworks (A. Tolba et. al, 2019). Due to the characteristics of the VANET, the moving vehicles are connecting and sharing information in specific secure infrastructures. For example, when one vehicle directly connecting and sharing information to another vehicle is known as Vehicle to Vehicle (V2V) and with the help of the Road Side Unit (RSU) when vehicle directly connecting and sharing information to infrastructure is called Vehicle-to-Infrastructure (V2I) (Z. Zhou et. al, 2017).

The Internet of Things (IoT) deals with the sensor embedded hardware and other digital devices interfaces with API for sharing the information among the nodes using internet. When the implementation of the framework in such way that where vehicles are connected to internet and functions like ad-hoc network is called Internet of Vehicles (IoV). In the greatest transformation of the technological landscape in the wireless and mobile communications industries the characteristics of the IoT play quite significant role for designing and developing the smart vehicle (Z. Ning et. al, 2017). Also, the VANET enables the different types of facilities to the IoV which is able to easily deal with road transport odds entities. For better traffic flow control and secure transmission of information regarding cities-based technologies the systematic solution can be achieved with IoV.

Due to IoT, the smartness and intelligences has increased very fast and drawn the attention of the various industries to adapt such technology where as IoV network is one of them. For the automobile industries, the IoV is very emerging field, for better traffic flow control and secure transmission of information, because the IoT make it possible the concept of the smart cities. It is a dispersed network that provides for the use of linked vehicles and the data created by the VANETs (J. Wan et. al, 2016). An increase in the number of drivers and manual traffic control system leads to increase in the death rate by accidents. Therefore, to overcome the accidents rate and enable the facilities for the vehicles to share the information in real-time is the primary objective of the IoV (J. Chen et. al, 2017).

*Figure 1. Components to define the properties of the VANETs*

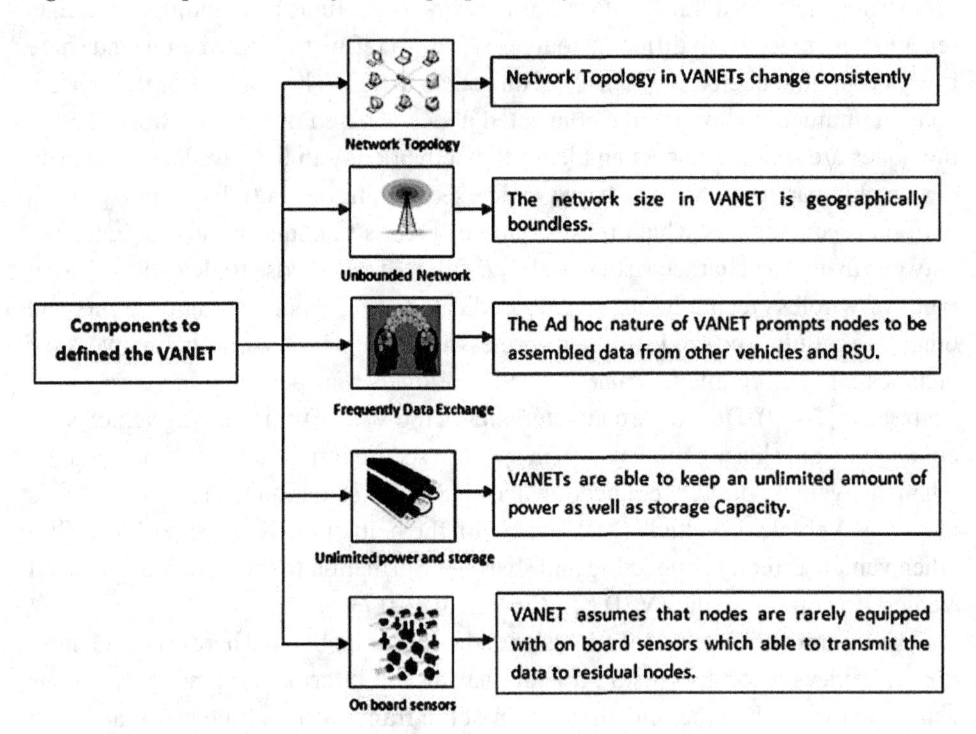


## RELATED WORKS

(Anand Paul et al., 2015) proposed an auxiliary perceptual intelligence for the Internet of Vehicles. The primary concern of the proposed model is that needed high wireless strength whereas it is lacking in computational resources and low bandwidth results shown as contradiction among the robust vehicular communication. For overcoming that concern, it needed resourceful cognitive radio along with efficient spectrum system. It has employed to reduce the problems of high vehicular mobility and lack of spectrum.

(Jiawen Kang et al., 2016) proposed two computer-generated mechanisms for designing and securing against failure for IoV. The primary function of the designed model is to identify the hidden location and draw using two virtual machines. The privacy order communication between a local and vehicles must be secure also facilitate to less usage of the model rules is the primary function of the active topology.

(Eun-Kyu Lee et al., 2016) introduced IoV named as self-guided edge-based vehicles using the concept of the smart grids. The recommendation of the researchers is that the cloud-based vehicle model is equally treated as internet-based cloud infrastructure with respect to the vehicles. Moreover, it would be quite robust framework that makes further enhancement and the self-directed manipulation of cloud planning will be the main achievement. In addition, the future scope has been suggested that the greater use of the unmanned aerial vehicles (UAVs) mechanism to design the cloud-based vehicle framework.

(Wenyu Zhang et al., 2017) proposed an IoV model which is based on the integration of fog computing and big data. The primary concept for developing the local IoV with less latency communication facilities because the problem get stated in cloud-based IoV networks. In addition, it has suggested that enhanced resource management mechanism which helps to advances the IoV framework such as intra-edge energy-efficient and inter-age QoS-efficient. Finally, the simulation outcomes of the proposed model get validated empirically.

(Wenchao Xu et al., 2017) made the study about Internet of Vehicles (IoV) and big data and get identified some of the major characteristics between both to integrate for developing the robust IoV model especially self-running vehicles. Moreover, after getting identifies the major issues and challenges in IoV there are certain necessary precautions has been suggested while designing the IoV with respect to big data.

(Xiaojie Wang et al., 2018) designed a framework which enables to taking off from real-time traffic management based on the edge computing IoV to minimize the average execution time and cost of the framework. Because in the modern digital era the technology is directing towards the artificial intelligence and machine leaning where the machine get trained as smart and intelligent but the performance always relies on the cost and efficiency. For traffic management systems, the utilization of a range of road side units communication to remove the information from fog nodes in terms of external vehicles did not get recognized while designing the framework.

(L. Sumi et al., 2018) proposed an IoT based VANET framework for controlling the traffic to assist emergency vehicles in an urban smart city. The framework is the intelligent use of IoT and VANET which make it easier passage through traffic paths for pre-existing emergency vehicles. Moreover, it assists to the emergency vehicles for finding the best path to reach at destination smoothly on the basis of real-time traffic data. In addition, it reduces the latency in transmission in case of medical emergency.

(M. A. Saleem et al., 2019) have introduced the congestion control framework for transmitting the information using VANETs based on the IoT. Where, it has used the fundamental concept of the routing protocols for VANETs. For transmitting the information among the vehicles or RSUs it needed huge amount of energy. The primary concept for finding the ideal path using the path finding techniques in VANETs gets done by the intermediate system which is efficiently manage the congestions. Conclusively, for transferring the information IoT based with VANET to efficiently overcome the congestion using fusion of clustering-based routing protocols.

## VANETS ARCHITECTURE

The architectural design of heterogeneous infrastructure with different techniques is a challenging task that requires identifying and effectively clustering the element-set with similar functionality and representation. The VANETs has basically divided into three domains (X. Hu et al., 2019) are following as:

### Mobile Domain

In mobile domain, it is basically consists of two parts such as vehicle domain and mobile domain. In vehicle domain, it includes all those vehicles which are continuously moving like buses, cars, trucks etc. In mobile device domain, it includes all portable handy devices like PDA, Laptop, GPS, Smart phone etc.

### Infrastructure Domain

The infrastructure domain also consists of two parts such as roadside infrastructure domain and central infrastructure domain. The domain of roadside infrastructure includes stationary roadside entities such as traffic lights, poles, etc. Whereas, the central infrastructure domain includes central management centers such as vehicle management centers, traffic management centers, etc.

### Common Domain

The common domain consists of internet infrastructure and private infrastructure. For example, various nodes and servers and other computing resources working directly or indirectly for VANETs fall under the common domain.

*Figure 2. Layered Architecture*

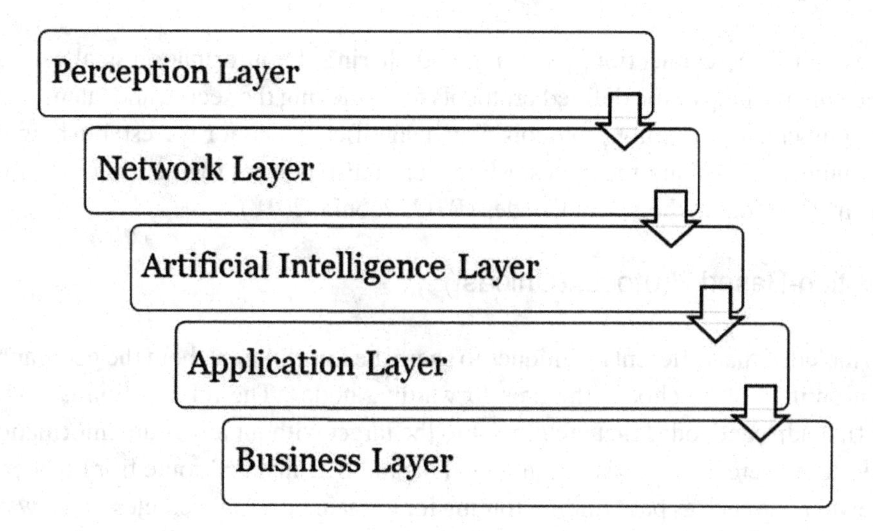


Another form of VANET architecture is communication architecture which is further classified (J. Toutouh et al., 2018) below:

## Vehicle to Vehicle Communication (V2V)

It is basically deals with the inter-vehicle communication. The prior connecting and sharing the information among the vehicles which is interconnect with the technique known as point to point architecture. It is facilitates with the characteristics where vehicular navigation becomes quite smooth and protective.

## Vehicle to Infrastructure Communication (V2C)

There are number of base stations located adjacent to a certain infrastructure for highways is required to enable with the services either uploading or downloading of data vice-versa vehicle also connecting and sharing the either to or from vehicles. The cluster gets covered to access point from each infrastructure.

## Cluster to Cluster Communication (C2C)

In terms of VANETs, the network is divided into groups which are defined as self-operated groups of vehicles. The communication gets established among the group of vehicles with the help of the base station manager agent. Moreover, it also facilitates with the services where one cluster can establish the connection with other clusters.

## Routing Protocols

To establish the connection, receiving and sharing the information is always get done with the help of the defined protocols for achieving the secure and latency-less environment. The routing protocols play a significant role for the establishing the communication among the nodes. Moreover, it also defines the distribution of the information among the various nodes (R. C. Poonia, 2018).

### Position-Based Protocol (Unicast)

It is included that different techniques to share the information about the geographic location in order to choose the next forwarding nodes. The information is sent to the first adjacent nodes that are closest to the target without any route information. Therefore, there is no need to require and maintain a universal route from sender to receiver node such as path finding routing for unmanned aerial vehicles in software-based networks (R. Yarinezhad et al., 2019).The routing protocols based on the rank are fundamentally classified into two classes such as location-based greedy vehicle-vehicle networking and delay-tolerant networking (DTN) protocols.

### Topology-Based Protocol

For the distribution of the packets to make it available the correct path for sharing the information is known as topology-based protocol. It is fundamentally distinguished with the help of the techniques named as active, reactive and hybrid. There is proactive table-based routing, where every hop maintains a routing table. In the reactive approach, it is on-demand routing protocol where there is no need to maintain routing tables. Lastly in the Hybrid routing, it is integration of both such as active and reactive where the mesh has classified into two domains (S. Glass et al., 2017).

### Relay-Based Protocol

The relay-based routing protocol is widely employed in vehicular navigation to share data regarding the display the notice and intimations between vehicles in case of emergency and traffic conditions as well.

### Cluster-Based Protocol (Multicast)

There is a group of hops which itself defines as the member of a group and any of the hope can be selected with the certain policies as cluster head i.e. assists to transfer the data from source to destination is called cluster-based protocol. When the network get clustered the outcomes as the latency rate will increased for the most vehicles but the scope of adaptability, where clustering needed, can be given in huge infrastructure (L. Tuyisenge et al., 2018).

### Specialized Multicast-Based Protocol

The objective of this protocol is to send the packets with the help of single vehicle through multiple vehicles to their respective destination based on their local geographic region labeled as domain. In unicast, it is facilitates to send the data within network of defined destination. But in specialized multicast, it is facilitates to transfer the data within network as well as outside network. The primary disadvantage of this protocol is network splitting as well as next adversarial nodes, which can disrupt normal information transfer (J. Contreras-Castillo et al., 2017).

## Security Measures in VANETs

The unsecured sharing of information using VANET communication can never be accepted because the sends and receive information must be precise, well ordered, and trustworthy. Each function of the VANET network is aims to facilitate the well ordered road safety with the help of the quick information sharing among the nodes. Otherwise, the single exploitation of the threats causes the loss of societal, economical and living life. The primary objectives for integrating the IoT in VANET system to design the secure framework where enhanced transportation safety, collision avoidance can be insured (O. Senouci et al., 2019).

- ***Confidentiality:*** The principle of this security measure is to maintain from unauthorized accesses of the information means only authorized person can see or read the information. Moreover, it is needed in group communication where team members are restricted to read information which is not belonging to them.
- ***Integrity:*** This security measure is to restrict the manipulation of the information like the data send from sender side should be identically received at receiver side. Although it is needed to change the information that only can be done by the authorized users. The techniques have used by the sender to encrypt the data that same technique get used to decrypt the data which is ensure the integrity of the data.
- ***Availability:*** It can be defined as when the authorized users make request for any resources it should be available. Similarly for the moving vehicles when the information get request by the vehicle it should be available in real-time which ensures the availability. The latency would not be acceptable if

maintaining the availability. While accessing the information from sensor-based system or ad-hoc network the response time should be efficient. Otherwise, the information get received to the vehicles little bit delay it should be useless for the system.

- ***Authentication:*** The authentication facilitates the authenticity of the messages as well as the sources. After receiving the information the behavior of the VANET nodes are acting same as expected which shows the legitimacy about the source and users.
- ***Reliability:*** The information received for communication must be correct and authentic. The unauthentic information gets terminated with the help of the random verification process.
- ***Anonymity:*** Usually the services get interrupted while driving the vehicles in unknown or remote locations. In such cases the security of the nodes has to be maintained using security measure. Moreover, it is not only ensure the security of the sensitive information but also available the services without any interruption.

## Security Threat in VANETs

In ad-hoc environments especially in the area of vehicles there are n numbers of attacks are possible. The purpose of such attacks is to gain the access of entire system through perform the any suspicious activity or make it fails the system. Attackers may engage in malicious activities with multiple intensions such as to take advantage of system features that is not authorized for that to make publish the sensitive information publically or some other purpose (J. Contreras-Castillo et al., 2017).

*Figure 3. Classical Categorization of the Attackers*

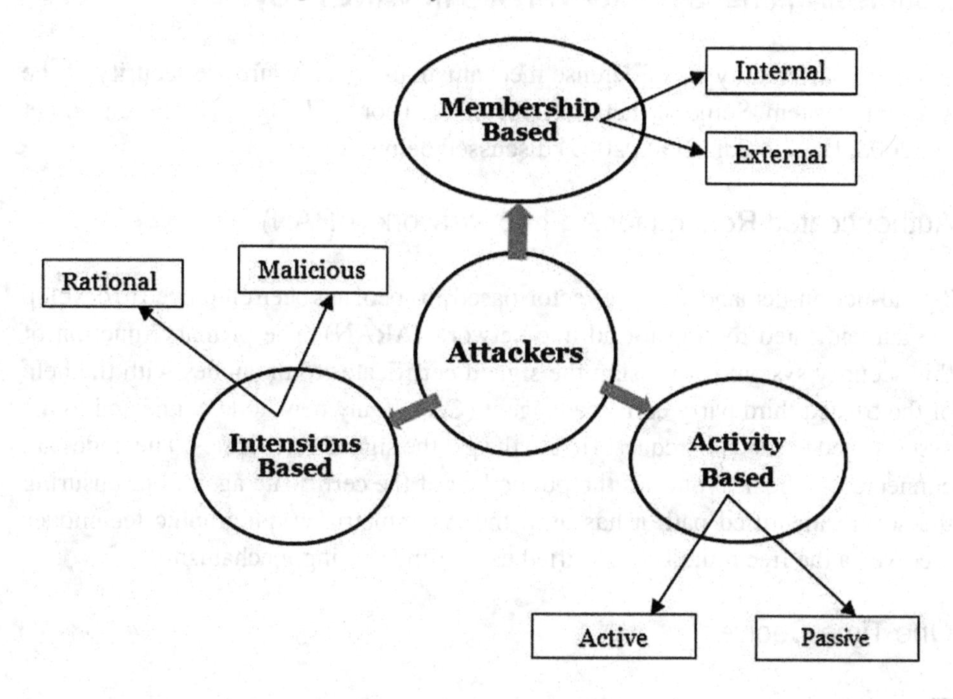


For the VANET systems the challenges are quite difficult because it contains very sensitive information. Therefore, it is very necessary condition to always keep the security as a primary concern while designing such types of the system which is sharing the sensitive information in real-time. Moreover, the developed system has the greater capability to tolerate the threats without getting impact of system.

*Figure 4. Classes of Attacks*

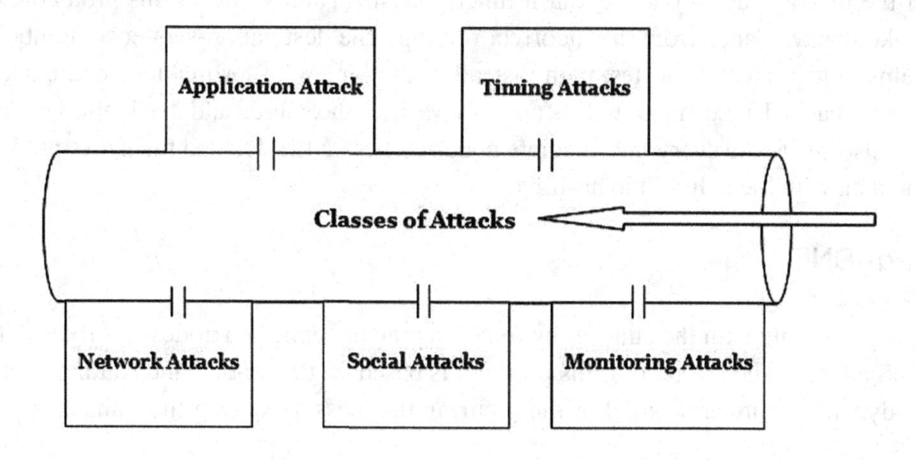

## Defense Against Security Threats in VANETs System

There are different types of defense mechanism uses for ensure the security of the VANETs system. Some of them have been (Y. Toor et al. 2008, H. Hartenstein et al., 2008, H. Moustafa et al., 2009) discussed below:

### Authenticated Routing for Ad-hoc Network (ARAN)

The ad-hoc on-demand distance vector-based protocol has been employed to develop the authenticated routing for ad-hoc network (ARAN). The primary function of this security system is to assign the signed certificates to the nodes with the help of the trusted third party certificate agent (CA). If any new node connected to the system need to send the request first to the get the signed certificates. The nodes are connected in the network has the public key of the certificate agent. For ensuring the secure identified path it has used the Asymmetric cryptographic techniques vice-versa the free route get identified using time sensing mechanism.

### One Time Cookie

The cookies are very important for creating a secure environment especially in session management where it gets assigned per session. Moreover, the one-time cookie has used to secure the framework from the session hijacking and theft of session ID. For every request the one-time cookie (OTC) has been generated.

### Secure and Efficient Ad-hoc Distance Vector Protocol (SEAD)

It is work over the destination-sequenced distance-vector routing (DSDV) protocol. For performing the authentication process to get verified the source and legitimacy of the message using one way hash function. The primary aim of this protocol is make the avoidance from the incorrect routing. The destination-sequence number helps to identify the shortest path instead of the long path for making secure and less to reached at destination. It is not only verifies the source and destination node but also verify the every intermediate nodes comes on the decided the path to send the data with the help of the hashing.

### ARIADNE

It is preventing with the attacker by removing the unconnected nodes and routes of DoS attacks. The concept of this protocol is based on the on-demand routing such as dynamic source routing. For the securing the message during the transmission it efficiently uses the symmetric cryptography. Using MAC, the sender to receiver and receiver to sender agrees on two keys at the sender and receiver both end.

## Secure Ad-hoc On-Demand Distance Vector Routing Protocol (SAODV)

When the security criteria get incorporated with AODV protocol is called the SAODV. All the coming and going messages are digitally signed to insure authenticity and the hop count get protected with hash function. For the security purpose the intermediate nodes are not involve in the any types of communication even though the nodes have the information about the shortest and secure path. Such issues can be resolve with the help of the double signature but the execution time gets increased.

## Adaptive-Secure Ad-hoc On-Demand Distance Vector Routing Protocol (A-SAODV)

It is an advancement of the SAODV with the additional characteristics along with the adaptive response judgment. The characteristics of this protocol gives the permission to the intermediate nodes involve in the communication to update the sender about best free available path based on the queue length and threshold conditions. Moreover, the issue of the SAODV gets resolved without increment in complexity.

## Elliptical Curve Digital Signature Algorithm (ECDSA)

The name itself defines about this protocol which is based on the digital signature. For making the secure environment there are two fundamental techniques used such as hash function and asymmetric cryptographic. The primary condition of this protocol is to sender and receiver has to follow the same terms and condition on the elliptical curve domain components.

## Integrated Protocol

While registering the vehicles with the help of the RSUs the authentication process get done with this protocol. In the first step of the vehicle registration, the vehicle needs to send the defined messages defined by the system to RSU and then RSU generates and reply back to vehicle with registration id and step by step authentication get done with the help of the certificate generated by RSU. Once the node get verified, it has eligible to shares the data otherwise it get blocked.

## CONCLUSION

The VANET intelligently incorporated with the IoT is very emerging wireless communication system for ensuring the highway safety and information services whereas the security is a primary concern. Because either without or less security can be made the huge loss in terms of life, societal and economical. Nevertheless, there are several challenges further which demonstrate the beneficial effect of VANET on traffic safety and efficiency. Now the modern smart cities need the smart and intelligent transportation management system where the vehicle can securely and easily connect with system and freely share the information. With the help of such ideas it can make the robust and establish the strong communication among the vehicle-to-vehicle, vehicle-to-infrastructure, and Vehicle-to-network. Finally, there are several defense mechanism has been suggested against the different types of attacks. The smart and intelligent incorporation of the defense mechanism can helps to design the robust framework.

## REFERENCES

Chen, J., Zhou, H., Zhang, N., Xu, W., Yu, Q., Gui, L., & Shen, X. (2017). Service-oriented dynamic connection management for software-defined internet of vehicles. *IEEE Transactions on Intelligent Transportation Systems*, *18*(10), 2826–2837. doi:10.1109/TITS.2017.2705978

Contreras-Castillo, J., Zeadally, S., & Guerrero Ibáñez, J. A. (2017a). A seven-layered model architecture for internet of vehicles. *Journal of Information and Telecommunication*, *1*(1), 4–22. doi:10.1080/24751839.2017.1295601

Contreras-Castillo, J., Zeadally, S., & Guerrero-Ibañez, J. A. (2017b). Internet of vehicles: Architecture, protocols, and security. *IEEE Internet of Things Journal*, *5*(5), 3701–3709. doi:10.1109/JIOT.2017.2690902

Feiri, M., Petit, J., Schmidt, R. K., & Kargl, F. (2013). The impact of security on cooperative awareness in VANET. *Vehicular Networking Conference (VNC)*, 127-134. 10.1109/VNC.2013.6737599

Glass, S., Mahgoub, I., & Rathod, M. (2017). Leveraging manet-based cooperative cache discovery techniques in vanets: A survey and analysis. *IEEE Communications Surveys and Tutorials*, *19*(4), 2640–2661. doi:10.1109/COMST.2017.2707926

Hartenstein, H., & Laberteaux, K. P. (2008). A tutorial survey on vehicular Ad Hoc Networks. *IEEE Communications Magazine*, *46*(June), 164–171. doi:10.1109/MCOM.2008.4539481

Hu, X., Wu, T., & Wang, Y. (2019). Social coalition-based v2v broadcasting optimization algorithm in vanets. In *International Conference on Swarm Intelligence*. Springer. 10.1007/978-3-030-26354-6_32

Kang, J., Yu, R., Huang, X., Jonsson, M., Bogucka, H., Gjessing, S., & Zhang, Y. (2016). Location privacy attacks and defenses in cloud-enabled internet of vehicles. *IEEE Wireless Communications*, *23*(5), 52–59. doi:10.1109/MWC.2016.7721742

Lee, E.-K., Gerla, M., Pau, G., Lee, U., & Lim, J.-H. (2016). Internet of vehicles: From intelligent grid to autonomous cars and vehicular fogs. *International Journal of Distributed Sensor Networks*, *12*(9). doi:10.1177/1550147716665500

Moustafa, H., & Zhang, Y. (2009). *Vehicular networks: Techniques, Standards, and Applications*. CRC Press.

Ning, Z., Hu, X., Chen, Z., Zhou, M., Hu, B., Cheng, J., & Obaidat, M. S. (2017). A cooperative quality-aware service access system for social internet of vehicles. *IEEE Internet of Things Journal*, *5*(4), 2506–2517. doi:10.1109/JIOT.2017.2764259

Paul, A., Daniel, A., Ahmad, A., & Rho, S. (2015). Cooperative cognitive intelligence for internet of vehicles. *IEEE Systems Journal*, *11*(3), 1249–1258. doi:10.1109/JSYST.2015.2411856

Poonia, R. C. (2018). A performance evaluation of routing protocols for vehicular ad hoc networks with swarm intelligence. *International Journal of System Assurance Engineering and Management*, *9*(4), 830–835. doi:10.100713198-017-0661-1

Saleem, M. A., Shijie, Z., & Sharif, A. (2019). Data transmission using iot in vehicular ad-hoc networks in smart city congestion. *Mobile Networks and Applications*, *24*(1), 248–258. doi:10.100711036-018-1205-x

Senouci, O., Aliouat, Z., & Harous, S. (2019). A review of routing protocols in internet of vehicles and their challenges. *Sensor Review*, *39*(1), 58–70. doi:10.1108/SR-08-2017-0168

Sumi, L., & Ranga, V. (2018). *An iot-vanet-based traffic management system for emergency vehicles in a smart city. In Recent Findings in Intelligent Computing Techniques*. Springer.

Tolba, A. (2019). Content accessibility preference approach for improving service optimality in internet of vehicles. *Computer Networks*, *152*, 78–86. doi:10.1016/j.comnet.2019.01.038

Toor, Miihlethaler, Laouiti, & De La Fortelle. (2008). Vehicle Ad Hoc Networks: Applications and Related Technical issues. *IEEE Communications Surveys & Tutorials, 10*(3), 74-88.

Toutouh, J., & Alba, E. (2018). A swarm algorithm for collaborative traffic in vehicular networks. *Vehicular Communications*, *12*, 127–137. doi:10.1016/j.vehcom.2018.04.003

Tuyisenge, L., Ayaida, M., Tohme, S., & Afilal, L.-E. (2018). Network architectures in internet of vehicles (iov): Review, protocols analysis, challenges and issues. In *International Conference on Internet of Vehicles*. Springer. 10.1007/978-3-030-05081-8_1

Wan, J., Liu, J., Shao, Z., Vasilakos, A. V., Imran, M., & Zhou, K. (2016). Mobile crowd sensing for traffic prediction in internet of vehicles. *Sensors (Basel)*, *16*(1), 88. doi:10.339016010088 PMID:26761013

Wang, X., Ning, Z., & Wang, L. (2018). Offloading in internet of vehicles: A fog-enabled real-time traffic management system. *IEEE Transactions on Industrial Informatics*, *14*(10), 4568–4578. doi:10.1109/TII.2018.2816590

Xu, Zhou, Cheng, Lyu, Shi, Chen, & Shen. (2017). Internet of vehicles in big data era. *IEEE/CAA Journal of Automatica Sinica, 5*(1), 19–35.

Yarinezhad, R., & Sarabi, A. (2019). A new routing algorithm for vehicular ad-hoc networks based on glowworm swarm optimization algorithm. *Journal of Artificial Intelligence and Data Mining*, *7*(1), 69–76.

Zhang, W., Zhang, Z., & Chao, H.-C. (2017). Cooperative fog computing for dealing with big data in the internet of vehicles: Architecture and hierarchical resource management. *IEEE Communications Magazine*, *55*(12), 60–67. doi:10.1109/MCOM.2017.1700208

Zhou, Z., Gao, C., Xu, C., Zhang, Y., Mumtaz, S., & Rodriguez, J. (2017). Social big-data-based content dissemination in internet of vehicles. *IEEE Transactions on Industrial Informatics*, *14*(2), 768–777. doi:10.1109/TII.2017.2733001

# Chapter 9
# Temporal Blockchains for Intelligent Transportation Management and Autonomous Vehicle Support in the Internet of Vehicles

**Zouhaier Brahmia**
https://orcid.org/0000-0003-0577-1763
*University of Sfax, Tunisia*

**Fabio Grandi**
https://orcid.org/0000-0002-5780-8794
*University of Bologna, Italy*

**Rafik Bouaziz**
https://orcid.org/0000-0001-5398-462X
*University of Sfax, Tunisia*

## ABSTRACT

*In the internet of vehicles (IoV) field, blockchain technology has been proposed for durable and trustworthy bookkeeping of the exchanged data. However, block timestamps assigned by miners are usually delayed with respect to events that generate the stored data, making them unusable for applications dealing with exact timing, like traffic law enforcement and insurance accident investigation. To overcome this shortcoming, the authors propose to add new timestamps to the blockchain, which are assigned by data originators to represent the valid time of data recorded within a transaction. The resulting enhanced blockchain data model, named BiTchain, can*

DOI: 10.4018/978-1-6684-3610-3.ch009

*be considered from a temporal database perspective as a bitemporal data model. In order to let users and applications enjoy the potential of BiTchain, they also introduce an expressive temporal query language, named BiTEQL, defined as a TSQL2-like temporal extension of the EQL blockchain query language.*

## INTRODUCTION

Present and future intelligent transportation and autonomous vehicles applications are using Internet of Vehicles (IoV) (Yang et al., 2014), Flying Ad-Hoc Networks (FANETs) (Bekmezci et al., 2013) and Vehicular Ad-Hoc Networks (VANETs) (Zeadally et al., 2012) as communication infrastructures. In this context, the Blockchain technology has been proposed for durable and trustworthy bookkeeping of the exchanged data (Guo et al., 2018; Diallo et al., 2020; Fu et al., 2020; Gupta et al., 2020; Li et al., 2020a; Narbayeva et al., 2020; Rehman et al., 2020; Javaid et al., 2021; Uddin et al., 2021; Chondrogiannis et al., 2022; Six et al., 2022). A blockchain (Nofer et al., 2017; Dinh et al., 2018; Zheng et al., 2018; Li et al., 2020b) is a public distributed ledger that stores committed transactions in an ordered list, or a chain, of blocks. Such a ledger is maintained by multiple distributed nodes (computers), which are linked in a peer-to-peer network (i.e., without a server node that has full control and central authority) and possibly do not trust each other, through a consensus mechanism (i.e., an agreement among these nodes on the truth of data stored in the blockchain) and cryptography (essentially hash algorithms and digital signatures). Notice that the ledger is replicated over all nodes. A transaction is a sequence of operations applied on some state respecting the ACID properties as in classical database systems. It cannot be modified once it is recorded in a block of the blockchain. A block is made up of a block header and a block body: the former contains, among others, the parent block hash that points to the previous block and a timestamp; the latter essentially contains transactions.

From a database point of view, a blockchain (Dinh et al., 2018; Mohan, 2018; Vo et al., 2018; Xu et al., 2019) can be considered as a distributed and replicated temporal database (Grandi, 2015; Jensen & Snodgrass, 2018a) that stores a large list of blocks: each timestamped block is replicated over all nodes of the network. More precisely, since block timestamps cannot be updated and the whole blockchain is an append-only data structure, a blockchain can be viewed as a transaction-time (Jensen & Snodgrass, 2018b) database, that is a database where each version of a time-varying datum (like the salary of an employee or the price of a product) is stamped with the time when such a version has been inserted in the database.

Among the strengths of blockchain, we find immutability, data integrity, transparency, distribution, absence of central/intermediary entity, and provision of trust in trustless environments. These advantages have lead to a wide use of blockchain technology in several fields, like data science (Liu et al., 2020), supply chains (Behnke & Janssen, 2020), business (Frizzo-Barker et al., 2020), economics (Kher et al., 2020), medicine (Zhang & Boulos, 2020), Internet of Things (IoT) (Pavithran et al., 2020), smart contracts (Singh et al., 2020), e-voting (Khan et al., 2020), finance (Chang et al., 2020), reputation systems (Sharples & Domingue, 2016), and security services (Li et al., 2020b), although it was initially developed to manage cryptocurrencies as Bitcoin (Nakamoto, 2008) and its successors.

Recently, the deployment of blockchains has also been proposed for smart traffic management and autonomous vehicles support (Guo et al., 2018; Fu et al., 2020; Gupta et al., 2020; Narbayeva et al., 2020; Rehman et al., 2020). An autonomous vehicle (Anderson et al., 2014; Taeihagh & Lim, 2019) (also known as an automated or self-driving vehicle, or as a driverless, self-driving or robotic car) is a vehicle that can, without interaction with a human driver: (i) sense its environment (detecting objects like other vehicles, analyzing their behaviour and predicting their evolution) and (ii) efficiently drive itself along several types of roads having different characteristics and possibly presenting complex and unexpected situations. Exchange of information about the environment is also carried out via communication with other vehicles and roadside devices (e.g., smart grids, hotspots). The autonomous vehicle technology aims at having a safe and comfortable driving, avoiding road accidents or reducing their damage, improving mobility for disabled and elderly persons, reducing fuel consumption, resolving traffic congestion, and efficiently exploiting available infrastructures. It is closely related to several other technologies coming from different fields like computer science, electrical engineering, robotics, and controls. Hence, advances in these domains will have a tremendous impact on autonomous vehicle industry. Moreover, although a lot of progress has been made during the past four decades, several aspects remain not sufficiently studied, like the one that we will deal with in the rest of the chapter, and which is related to the use of blockchains for the management of these vehicles.

In fact, accurate timestamping of stored data (e.g., starting from vehicle trajectory tracking) is essential for traffic management and control. To this purpose, the basic blockchain timestamp support is not sufficient. Indeed, due to the mining effort, block timestamps are not accurate (Zhang et al., 2019b; Szalachowski, 2018; Ma et al., 2020; Zhang et al., 2019a) and at least a 10÷15-minute discrepancy with the events originating the data is usually considered. For example, in the Bitcoin blockchain (Nakamoto, 2008; Tschorsch & Scheuermann, 2016), the timestamp of a block[1] is required to be accurate only within two hours of the network-adjusted time. Some solutions were proposed to reduce such a discrepancy, like the Median Past Time Rule and Future Block Time Rule which are used to prevent timestamp abuse for Bitcoins[2].

Therefore, in addition to the timestamp in the block header, put there by the miner and which can be considered as a blockchain-related transaction time from a temporal database point of view (as explained above), we think that also a valid timestamp (Jensen & Snodgrass, 2018c) is needed within the data, put there by the data originator (corresponding to the accurate validity of the data, at least according to the local/network clock of the system that recorded the data), giving rise to the need for a bitemporal (Jensen & Snodgrass, 2018d) data model (i.e., a model that maintains the history of data stored in the blockchain along both transaction and valid time). This new bitemporal model that we will propose in this work is named BiTchain (Bitemporal Blockchain). Notice that in the temporal database literature (Grandi, 2015; Jensen & Snodgrass, 2018a), the valid time (Jensen & Snodgrass, 2018c) of a data version is the time when such a version is current in the modeled reality.

Existing query languages proposed for blockchains (e.g., EtherQL (Li et al., 2017) or EQL (Bragagnolo et al., 2018)) of course do not provide a built-in support for querying and modifying bitemporal data and, thus, in this work we will complete the figure of our proposal with a new temporal query language, named BiTEQL (BiTemporal EQL). BiTEQL is a TSQL2-like temporal extension of EQL for querying BiTchain blockchains. TSQL2 (Temporal SQL2) (Snodgrass et al., 1995) is a consensus temporal query language specified as a temporal extension of the SQL92 standard and designed by a committee of temporal database experts chaired by R. T. Snodgrass in 1995, and EQL (Ethereum Query Language) (Bragagnolo et al., 2018) is an existing SQL-like language for querying blockchain databases.

Hence, our contribution, which has been first introduced in (Brahmia et al., 2022), is twofold: First, a new bitemporal data model for blockchain representation in the traffic management and autonomous vehicles field. This model is an extension with valid time of the current blockchain model. Second, a new temporal query language, associated to the proposed model, for querying bitemporal blockchains. This language is an expressive temporal extension of the EQL language inspired by TSQL2.

The rest of this chapter is structured as follows. The next section describes the background of our work and the motivation behind our chapter. In "BiTchain: A New Data Model for Bitemporal Blockchain Management", we propose BiTchain, a data model for representing and managing bitemporal blockchains. The section "BiTEQL: A Temporal Query Language for Querying BiTchain Blockchains" introduces BiTEQL, a temporal query language for querying BiTchain databases, and illustrates its functioning with some examples. The "Related Work Discussion" section discusses related work. Finally, conclusion and some remarks about our future work are provided.

## MOTIVATION AND BACKGROUND

In this section, we briefly motivate our approach and illustrate the background framework of our contribution.

*Figure 1. A networked intelligent vehicle in the reference application scenario.*

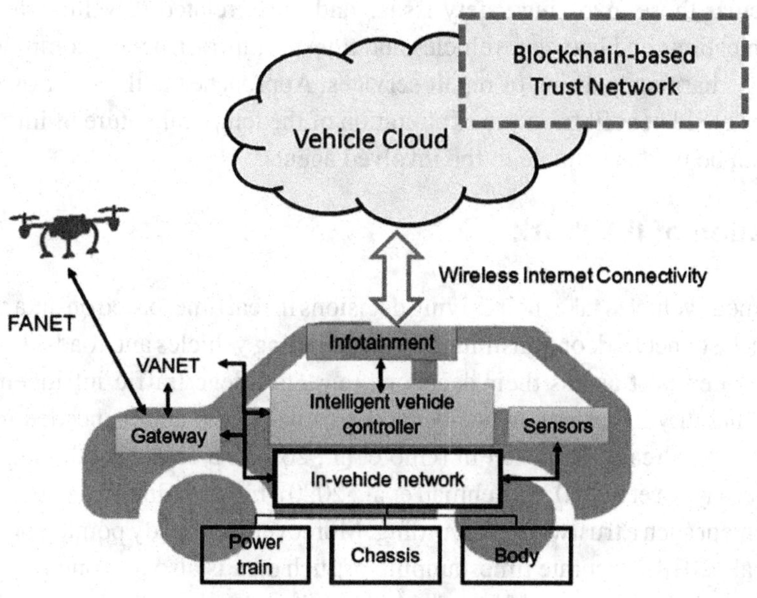


Among the most important IoT applications, we find intelligent transportation systems which aim at facilitating road transportation with unprecedented efficiency and safety. The communication infrastructure of an intelligent transportation system is a peer-to-peer mobile network (including a VANET and possibly a FANET) essentially encompassing smart or autonomous vehicles which directly communicate with each other and with roadside base stations, while almost continuously moving. Wireless internet connectivity enables the deployment of a vehicular cloud network for information sharing. The spearhead of intelligent transportation management can be considered the autonomous vehicle technology, which is still in its youth, although there are some interesting commercial solutions like the Waymo One[3] of Waymo LLC and the Volvo XC90 SUV of Uber[4], and semi-autonomous driving can be rather considered the state-of-the-art (e.g., Tesla cars equipped with the Autopilot feature[5]).

Thanks to its advantages (including distribution, absence of central parties, neutral support of security and reliability of transactions), the blockchain technology is being adopted for autonomous vehicles support[6] (Guo et al., 2018; Fu et al., 2020; Gupta et al., 2020; Narbayeva et al., 2020; Rehman et al., 2020). In particular, the adoption of a blockchain has been proposed to set up a trust network among intelligent connected vehicles (Kim, 2018; Li et al., 2021; Tyagi et al., 2022). With reference to the application scenario depicted in Figure. 1, we assume a blockchain-based trust network to be part of the vehicle cloud, in order to support secured advanced applications. However, in this context, some problems have not been resolved yet, in particular those involving safety issues and those related to vehicle-to-vehicle communications and between vehicles and transport infrastructure components for refueling, charging, parking or repair services. Approaches to the solution of these problems could benefit from the exploitation of the temporal nature of information which can be made available to the involved agents.

## Motivation of the Work

Autonomous vehicles take their driving decisions in real time, based on data acquired by their sensor network or transmitted by surrounding vehicles and roadside devices. In order to ex post assess their decisions (e.g., to allege traffic infringements or evaluate liability in case of an accident), a trustworthy system is needed to record such data. As already proposed in (Guo et al., 2018; Fu et al., 2020; Gupta et al., 2020; Narbayeva et al., 2020; Rehman et al., 2020), blockchains are an elective way to implement such a trustworthy recording. Moreover, as already pointed out, e.g., in (Guo et al., 2018), accurate timestamping of such data is also a strong requirement to reconstruct the premises of a real-time decision. Accurate timestamping is also needed for keeping track of real time actions performed for traffic control and safety enforcement by road control authorities and intelligent transportation support systems. With current blockchain technology indeed, the timestamp of a block, representing when the block was added to the blockchain, is rather approximate, as shown in (Zhang et al., 2019b; Szalachowski, 2018; Ma et al., 2020; Zhang et al., 2019a). For example, in Bitcoin and blockchain systems based on it, like Blockstack (Ali et al., 2016) and Catena (Tomescu & Devadas, 2017), the inaccuracy of block timestamps results from two causes (Zhang et al., 2019a):

- Due to the computational effort made by miners (i.e., participants who maintain the blockchain and add new blocks containing transactions), block timestamps may be erroneous up to two hours.
- Once it is committed, a transaction is not immediately added to the blockchain; it must wait for almost one hour before being integrated into a block and, thus, the transaction time of data does not correspond to the occurrence time of the event that generated the data.

Since "the hour is the hour; before the hour, it is not the hour; after the hour, it is no longer the hour" (as said by Jules Jouy), block timestamps are not very useful for any application that requires accurate timestamping of stored data, and, in particular, for intelligent transportation and autonomous vehicle applications. Therefore, in order to satisfy the requirements of these applications, we have to extend the existing blockchain data model with a new temporal property or valid timestamp that represents the existence time of data that are supplied by a data originator (i.e., a device which inputs data in the recording systems) and packaged inside blockchain transactions. This time is assigned by the data recording system, according to its local/network clock.

To this purpose, owing to a decades-long experience in temporal database research (Grandi, 2015; Jensen & Snodgrass, 2018a), we propose to extend a blockchain with valid time in order to enable the exploitation of a new bitemporal blockchain data model (BiTchain).

Since a blockchain is storing massive amounts of data, there is an increasing demand for effectively and efficiently querying blockchains in order to make their information contents available to applications by means of a powerful and user-friendly query language. Moreover, there is no consensual proposal or standardization effort concerning temporal query languages for blockchains, although they can be considered temporal (and more precisely, transaction-time) databases. Therefore, taking into account these remarks and in order to allow blockchain users and applications to efficiently exploit our BiTchain model, we decided to introduce a suitable temporal query language (BiTEQL) inspired by the well-known consensual language TSQL2.

Figure 2 presents an example that illustrates the problem that we deal with in this chapter and the solution that we propose for such a problem. The example involves five events in sequence (for the sake of simplicity, named A to E) and shows how data concerning an event submitted at the occurrence time of the event can be persistently recorded in a standard blockchain with a different transaction time. Let us consider the data which follow:

- data concerning event A occurred at time 1, recorded in the blockchain at time 13;
- data concerning event B occurred at time 3, recorded in the blockchain at time 9;
- data concerning event C occurred at time 5, recorded in the blockchain at time 7;
- data concerning event D occurred at time 6, recorded in the blockchain at time 14;
- data concerning event E occurred at time 9, recorded in the blockchain at time 10.

The resulting blockchain is shown in the top part of the figure, where the transaction timestamps do not provide any useful timing of the events concerning the recorded data. This can be only derived from valid time as represented in the timeline in the bottom part of the figure. Only the addition of valid time allows users and applications to verify, for instance, that event A preceded event C (by 4 seconds, assuming 1 second as the time unit in the example).

Notice that details on the blockchain structure and on our valid-time extension will be found later.

*Figure 2. A motivating example: the transaction time of blocks can be definitely useless to reconstruct the sequence of events and their exact timing.*

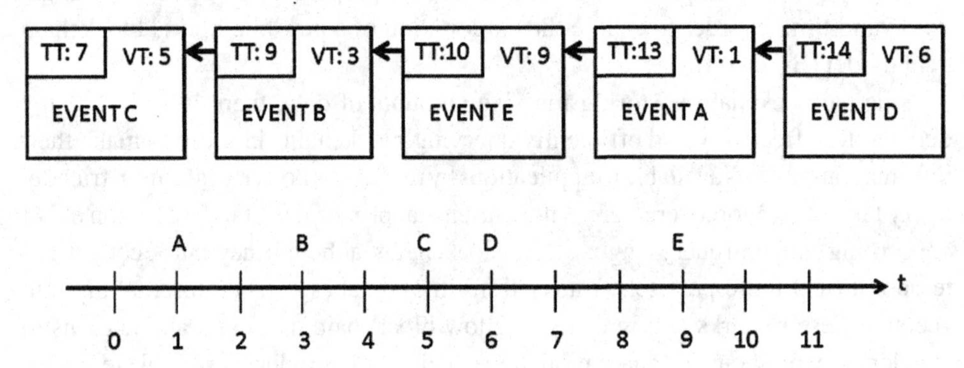


## Background on Temporal Databases

A temporal database is a database with built-in support for managing time-varying data (Grandi, 2015). Since the 1980s, temporal databases have attracted the interest of database researchers, vendors and practitioners. Hundreds of papers have been written on different aspects of temporal databases, as shown by several bibliographies: the first one (Bolour et al., 1982) was published in 1982, and the last one (Grandi, 2012) was published in 2012.

Temporal extensions to data models or query languages consist of supporting one or more temporal dimensions, according to the applications and users' requirements. Transaction time (i.e., the time automatically assigned by the system) and valid time (i.e., the time provided by the end user) are the two most important temporal dimensions that have been widely used and accepted by the temporal database community (Jensen et al., 1998). A database that supports only transaction time (valid time, resp.) is called a transaction-time (valid-time, resp.) database: the history of each time-varying data is kept only along transaction time (valid time, resp.), which means that each version of a time-varying data has only a transaction (a valid, resp.) timestamp. A database that supports both transaction time and valid time is called a bitemporal database: each version of a time-varying data has both a transaction timestamp and a valid timestamp. A database in which coexist data of different temporal formats (transaction-time, valid-time and bitemporal) is called a multitemporal database (Jensen et al., 1998).

In order to efficiently exploit temporal databases in applications, many temporal query languages have been defined; the most popular one is TSQL2 (Snodgrass et al., 1995) that has been proposed for querying bitemporal relational databases. The latest version of the SQL standard, SQL:2011, supports several concepts and constructs of TSQL2 (Kulkarni & Michels, 2012).

## Background on Traditional Data Model of Blockchain

A blockchain is a sequence of blocks. Figure 3 shows an example of a blockchain. Each block is composed of a header and a body and is linked to its parent block (i.e., its predecessor in the sequence) via a cryptographic hash pointer located in the block header. The first block of a blockchain is called "genesis block" and does not have a parent block. Figure 4 shows the inner structure of a block.

*Figure 3. An example of a blockchain.*

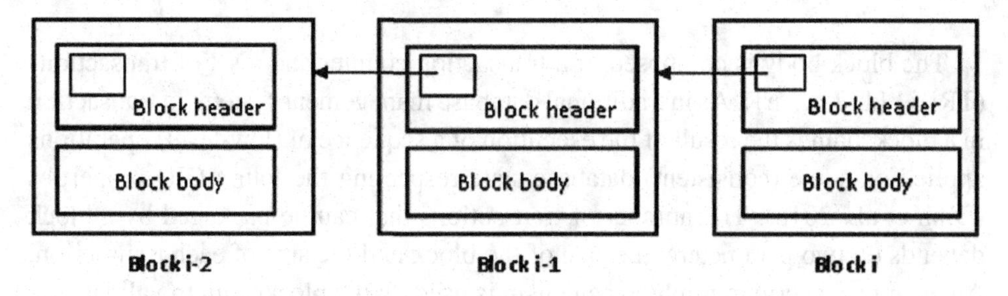


The block header contains six components (Zheng et al., 2018):

- Block version: the set of block validation rules to follow.
- Merkle tree root hash: the hash value of all transactions in the block.
- Timestamp: the current time as seconds in universal time, since January 1, 1970.
- nBits: the target threshold of a valid block hash.
- Nonce: a 4-byte field, usually starting with 0 and increasing for each hash calculation.

*Figure 4. The structure of a block.*

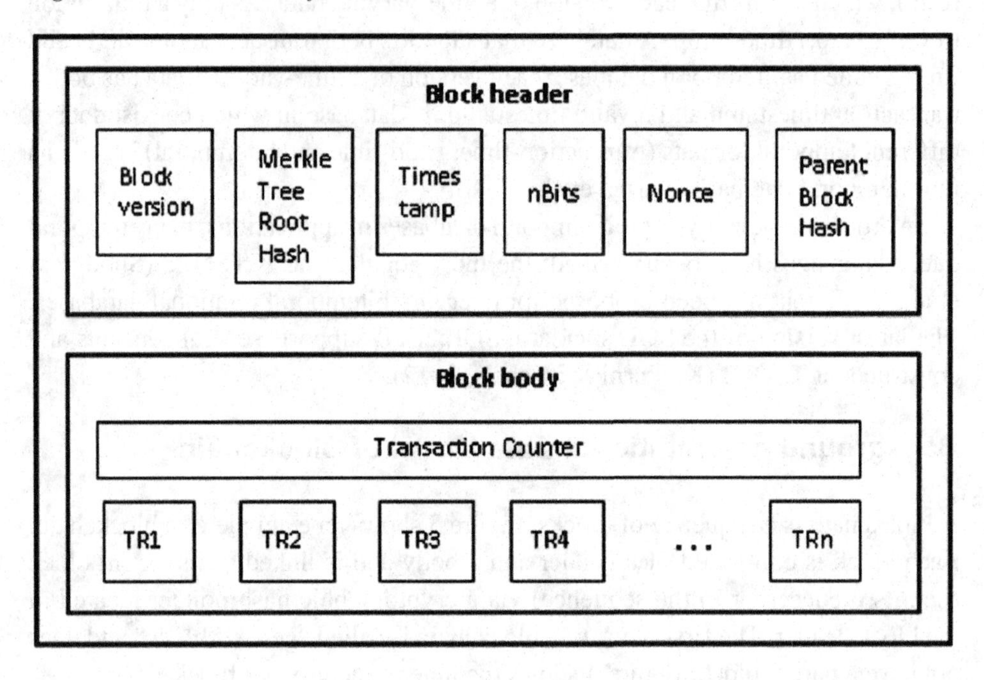


- Parent block hash: a 256-bit hash value that points to the previous block in the chain.

The block body is composed of a transaction counter and a set of transactions (TRi, i Î {1, 2, …, n}). As in traditional database management systems, a transaction in a blockchain is the result of the execution of a sequence of (low-level) operations applied on some (consistent) database state respecting the four ACID properties (Dinh et al., 2018). The number of transactions that can be packaged by a block depends on two parameters: the size of the block and the size of each transaction. An asymmetric cryptography mechanism is used by the blockchain to validate the transaction authentication.

As mentioned in previous sections, from a temporal database perspective, the traditional data model of a blockchain can be considered as a transaction-time data model.

## Background on the EQL Query Language

EQL is a SQL-like language which has been proposed for the Ethereum blockchain. Its "Select" statement is similar to that of SQL and includes the following five clauses: select, and from, which are mandatory, where order by and limit, which are optional. Moreover, EQL is case-insensitive (like SQL) and has no group by clause (unlike SQL).

The syntax and semantics of this language, with some query examples, could be found in (Bragagnolo et al., 2018).

We have chosen this language as a basis of our work for the following reasons:

- it is a SQL-like language and, thus, users or developers who are familiar with SQL will not find great difficulties in using EQL or a temporal extension of it;
- it allows blockchain users (a) to specify structural and semantic filters, (b) to reformat and transform returned data, (c) to order query results, and (d) to limit the number of returned blocks.

# BiTchain: A NEW DATA MODEL FOR BITEMPORAL BLOCKCHAIN MANAGEMENT

In this section, we introduce BiTchain, our bitemporal blockchain data model proposed in (Brahmia et al., 2022) that extends traditional blockchains to the valid time dimension support.

## Temporal Extension of the Traditional Data Model

The purpose of our proposed blockchain extension with valid time consists of allowing the data originators to specify, for each transaction, the time the data recording was really done, at least according to the local/network clock of the system that recorded the data. Such time only depends on the "will" of the data originator through the execution of the recording transaction and is not affected by the blockchain mining effort.

Therefore, we obtain a bitemporal blockchain data model that maintains the history of data along both transaction time (block timestamp, provided by the miner) and valid time (data validity timestamp, supplied by the data originator). All the sets of data produced by the transactions contained in each block have the same transaction timestamp, since there is only one block timestamp, whereas each set of data produced by the same transaction have the same valid timestamp, since there is a (different) recording timestamp per transaction (and, thus, several valid timestamps per block).

Thus, w.r.t. the traditional blockchain data model, as shown in Figure 4, the proposed valid timestamp will be associated as a new property named ValidTime to each transaction (TRi) packaged in the block body.

Hence, the first part of our contribution is the definition of the BiTchain data model, which follows. In the BiTchain data model, the blockchain BC can be regarded (and queried) as a bitemporal table in the TSQL2 data model: BC.transactions(A1, A2, …, Am | TT, VT), whose tuples correspond to the blockchain Transactions. Explicit attributes A1, A2, ..., Am are the transaction data components of the blockchain (which we assume for simplicity to be flat values in order to have a 1NF temporal relation) and implicit attributes TT and VT are the transaction timestamp extracted from the blockchain Block containing the Transaction, and the valid timestamp extracted from the blockchain Transaction property ValidTime, respectively. In particular, in the TSQL2 terminology, BC is always an event table w.r.t. transaction time, since TT timestamps are timepoints, but can be an event or state table w.r.t. valid time if valid timestamps added to the blockchain Transactions are timepoints or periods, respectively. In practice, the table BC plays the role of a relational view, and can be implemented as a SQL foreign data wrapper, encapsulating the underlying blockchain data as follows:

BC.Blocks.BlockBody.Transactions.Aj → BC.transactions.Aj (j ∈ {1, 2, …, m})
BC.Blocks.BlockBody.Transactions.ValidTime → VT(BC.transactions)
BC.Blocks.BlockHeader.Timestamp → TT(BC.transactions)

We also assume the BiTchain model to be a consistent extension of the EQL data model, that is the values of all the other components of the blockchain structure can be accessed from BC via explicit path expressions as defined in the EQL data model (Bragagnolo et al., 2018) (e.g., the properties Nonce, TransactionCounter, ParentBlockHash of the blockchain BC can be referenced in queries via EQL path expressions BC.blocks.header.nonce, BC.blocks.body.amountOfTransactions, BC.blocks.header.parentHash, resp.). They can be dealt with as additional explicit attributes as they are uniquely defined for each tuple of the BC.transactions bitemporal table.

## BiTEQL: A TEMPORAL QUERY LANGUAGE FOR QUERYING BiTchain BLOCKCHAINS

In order to efficiently exploit the potentialities of our BiTchain data model, we present in this section a temporal query language, first proposed in (Brahmia et al., 2022), named BiTEQL (Bitemporal EQL) defined as a TSQL2-like temporal extension of EQL. In practice, the syntax and semantics of BiTEQL are exactly the same as the syntax and semantics of TSQL2, extended with the specific EQL constructs to deal with blockchains (in the same way that EQL extends SQL).

In the following, first we describe time representation and manipulation facilities in BiTEQL, then we describe temporal selection and temporal projection support in BiTEQL, and finally we illustrate the BiTEQL expressive power through some sample queries.

## Time Representation and Manipulation in BiTEQL

As in the TSQL2 model, time is considered discrete, with a minimal system-dependent representation unit called chronon (Jensen et al., 1998). A mono-temporal chronon corresponds to an elementary interval on the time axis, whereas a bitemporal chronon corresponds to the Cartesian product of a transaction-time and a valid-time chronon. Three base temporal types have been defined for TSQL2 at the conceptual level: date-time, period and interval. The first one corresponds to an instantaneous event, without duration, which can be conventionally represented via a single chronon. The second one corresponds to a set of consecutive chronons along the time axis and is characterized by two datetime constants which represent its boundaries. The third corresponds to a pure duration, non-anchored on the time axis, and can be represented as a multiple of the chronon. Union of disjoint time periods are also called temporal elements.

Functions and operators defined to manipulate time and duration datatypes are assumed to be available. We further assume that, as it happens for TSQL2, casting from another temporal datatype to a duration can be used to calculate the overall duration of a time period or element by means of, for instance, type constructor functions.

The TSQL2 language, which is based on a bitemporal data model, provides functions to access the valid and transaction time timestamps, which are implicit attributes of tuples. We assume similar functions to be available also for BiTEQL: if T is a bitemporal tuple, the expressions **VALID**(T) and **TRANSACTION**(T) can be used to extract the valid and transaction timestamps of T, respectively.

## Temporal Selection in BiTEQL

Temporal selection is the most qualifying feature of a temporal query language, as it allows selecting data on the basis of their temporal properties. In order to add temporal selection capabilities to the EQL language, we extend the syntax of the **WHERE** clause of the **SELECT** statement.

As each transaction in a block is correspondingly augmented with the new valid timestamp in the data model, the BiTEQL **WHERE** clause is extended with variables and functions in order to specify temporal selection conditions based on data validity. In particular, in the BiTEQL **WHERE** clause, TSQL2 temporal (binary infix) predicates can be used, with the same semantics, to specify constraints over timestamp variables.

The available comparison operators are presented below with their semantics (the functions **BEGIN**(T) and **END**(T) can be used to extract the first and last chronon from T, respectively):

1. A **PRECEDES** B: **END**(A) is earlier than **BEGIN**(B).
2. A = B: A and B are identical (i.e., they contain the same chronons).
3. A **OVERLAPS** B: the intersection of A and B is not empty.
4. A **MEETS** B: **END**(A) immediately precedes **BEGIN**(B).
5. A **CONTAINS** B: each chronon in B is also contained in A.

These operators can be used to compare (monodimensional) temporal elements, periods and time points. Since all temporal types can be reduced to sets of chronons, such operators can also be used to compare operands with different temporal types (Snodgrass et al., 1995). For instance, if A is an element and B is a period, then the expression "A **PRECEDES** B" is true if the last chronon belonging to A precedes the left boundary of B. All the comparison operators can be implemented on the basis of a primitive operator "Before ()" which defines the relation order on the time axis. It can also be easily checked that such operators guarantee the temporal completeness of the resulting language, as they allow users to check the occurrence of all the possible relationships between two periods or events (Allen, 1983). Such operator set has been chosen for TSQL2 also considering the user-friendliness of the language among the design principle. This led to a non-minimal set of comparison operators which are closer to their meaning in natural language than the artificial definition of operators which equip other temporal languages (e.g., based on Allen's algebra (Allen, 1983)).

The **OVERLAPS** and **CONTAINS** operators can also be defined to work on multi-dimensional timestamps in a straightforward way.

## Temporal Projection in BiTEQL

Temporal projection is the operation which specifies the value of the timestamps to be assigned to the retrieved data. TSQL2 supports an optional **VALID** clause to specify valid-time projection, as the transaction time assigned to query results is always the current time and cannot be changed by the user. In case a **VALID** clause is not specified, the data timestamps are used by default (if more than one temporal relation R1, R2, …, Rn appear in the **FROM** clause, the valid time assigned to the result is computed by default as the intersection of the timestamps of the data in R1, R2, …, Rn in order to implement an implicit temporal join). BiTEQL also supports such a clause.

## Examples of BiTEQL Queries

In this section, we illustrate the use of BiTEQL by providing some examples of queries written in this language.

First of all, the basic EQL syntax of BiTEQL can be used to express standard queries over blockchains. For instance, Listings 1 below shows a BiTEQL query over blocks of the blockchain BC similar to that presented in Section 4.2 of (Bragagnolo et al., 2018):

*Listing 1. BiTEQL query over blocks*

```
SELECT block.parent.number, block.hash,
       block.timestamp, block.number,
       block.amountOfTransactions
FROM   BC.blocks AS block
WHERE TRANSACTION(block) = '2021-04-12'
  AND block.transactions.size > 10
ORDER BY block.transactions.size
LIMIT 100;
```

The only difference is in the temporal selection: whereas in EQL the transaction timestamp of a block is an explicit property that can be referenced via the path block. timestamp, in BiTEQL it is an implicit attribute that can be accessed through the **TRANSACTION**() function as for TSQL2.

Furthermore, in order to unleash the BiTEQL full expressive power, the TSQL2-like extensions can be used to query the transactions of a blockchain as if they were relations in a temporal database. For example, we assume to deal with a blockchain named traffic, whose transactions are used to track the position and speed of moving vehicles. Its schema is: traffic.transactions(VID, X, Y, VX, VY, NID | TT, VT), where VID is the vehicle identifier, X and Y are the coordinates of its position, VX and VY are the components of its velocity (TT and VT are the implicit timestamps). NID is the identifier of the communication network used for the acquisition of the vehicle data. We further assume that built-in functions are available to compute the scalar speed of a vehicle and the distance between two vehicles (or between a vehicle and a given point). If V1 and V2 are transactions stored in blocks of traffic, such functions can be defined as **SPEED**(V1) = $(V1.VX^2 + V1.VY^2)^{½}$ and **DISTANCE**(V1,V2) = $((V1.X-V2.X)^2 + (V1.Y-V2.Y)^2)^{½}$ . Then we can see some examples of TSQL2-like temporal queries issued with BiTEQL on the traffic blockchain.

Listing 2 shows a query doing a temporal selection exploiting the valid time of traffic transactions.

*Listing 2. BiTEQL query example*

```
SELECT VPOS.X, VPOS.Y, SPEED(VPOS)
FROM traffic.transactions AS VPOS,
WHERE VPOS.VID = 'XY794EU'
  AND VALID(VPOS) CONTAINS TIMESTAMP '2022-04-01 11:30';
```

This query retrieves the position and the speed of vehicle XY794EU at half past 11 in the morning on April 1, 2022.

The query which follows, displayed in Listing 3, exploits the nature of the bitemporal BiTchain model to show the efficiency of the communication networks used by different vehicles. Data concerning the traffic in 2021 are used for the analysis.

*Listing 3. BiTEQL query example*

```
SELECT NETID, AVG(TRANSACTION(VTRAN)- VALID(VTRAN))
FROM   traffic.transactions AS VTRAN
WHERE PERIOD '2021' CONTAINS VALID(VTRAN)
GROUP BY NETID;
```

In particular, the transactions stored in the traffic blockchain valid in 2021 are grouped by the network identifier NETID, and the difference between their transaction and valid times is averaged over each NETID group. Since the same blockchain is used for recording all the data, the delay introduced by miners equally effects on average all the NETID groups, provided that there is a sufficient number of transactions per group. Hence, the differences between delays computed by the query reflect the different behavior of the involved communication networks, as a greater average delay is introduced by a slower network.

One of the qualifying features of TSQL2, inherited by BiTEQL, is the capability to introduce in the FROM clause history variables (Grandi et al., 1993) to denote homogeneous temporal versions of the data objects stored in a relation. Hence

BiTEQL can be considered a history-oriented temporal SQL extension (Grandi et al., 1995), allowing users to write very expressive queries in a simple way. A simple use of such variables is shown in Listing 4, which exemplifies a combined use of TSQL2 and EQL constructs.

*Listing 4. BiTEQL query example*

```
SELECT TB.miner, TB.difficulty, SPEED(VTR)
FROM   traffic.blocks AS TB, TB.transactions AS VTR
WHERE VTR.VID = 'AR932TI'
  AND TRANSACTION(TB) - VALID(VTR) > 45 MINUTE
LIMIT 100;
```

In particular, for each transaction VTR concerning the vehicle AR932TI that required more than three quarters of an hour to be inserted into the blockchain, it retrieves (limiting the result to the first 100 tuples) the address of the miner having performed the insertion, the value of the corresponding blockchain difficulty parameter and the speed of the vehicle involved in the transaction (in order to see, e.g., whether there is some correlation between the delay of acquisition of the vehicle position data in the traffic blockchain and the speed of the vehicle, which could also be used as a quality indicator for the network communication behavior). Notice that the transaction time could also be extracted from VTR rather than from TB, as it is part of the (bitemporal) BiTchain schema adopted for traffic transactions.

Listing 5 exemplifies a more interesting use of history variables, where VPOS is declared to range over the traffic blockchain and V is declared to range over the VPOS transactions with the same value of VID, that is the transactions making up the whole history of the position and speed of the vehicle identified by VID.

*Listing 5. BiTEQL query example*

```
SELECT V.VID
FROM traffic.transactions AS VPOS, VPOS(VID) AS V
WHERE PERIOD '[2022-01-15]' CONTAINS VALID(V)
GROUP BY VALID(V) USING 1 MINUTE LEADING 9 MINUTE
HAVING AVG(WEIGHTED SPEED(V)) > 110;
```

Using a powerful temporal grouping mechanism defined for TSQL2, it computes at each minute the average speed of vehicles V over a sliding window of 10 minutes (preceding and containing the reference minute), during January 15, 2022. The modifier **WEIGHTED** is used within the aggregate function **AVG** in order to correctly evaluate the speeds with a weight given by their timestamp. Finally, identifiers of vehicles whose average speed exceeds the 110 Km/h limit are returned.

The next example can be used to highlight the importance of a bitemporal data model for the functioning of a traffic enforcement system using speed cameras and storing recorded data in a traffic blockchain. We assume two cameras C1 and C2 are installed along a route, along which a speed limit of 90 Km/h holds, at a precise distance of 10 kilometers from each other. Both cameras are able to identify vehicles via automatic plate recognition and record their passage time at the camera site in the traffic blockchain. The coordinate pairs (C1X, C1Y) and (C2X, C2Y) represent the location of the two camera sites, respectively (we use symbolic coordinates for simplicity).

*Listing 6. BiTEQL query example*

```
SELECT VPOS2.VID, VALID(VPOS2), 10*3600/(VALID(VPOS2) -
VALID(VPOS1))
FROM traffic.transactions AS VTRAN, VTRAN(VID) AS VPOS1 VPOS2
WHERE VPOS1.X = C1X AND VPOS1.Y = C1Y AND VPOS2.X = C2X AND
VPOS2.Y = C2Y
  AND 10*3600/(VALID(VPOS2)- VALID(VPOS1)) > 90;
```

Using history variables, the query in Listing 6 defines VPOS1 and VPOS2 as two versions of the same vehicle defined by VID. The **WHERE** clause ensures that the first (second) version corresponds to the passage of the vehicle at the position of the C1 (C2) camera site, and checks whether the speed limit has been broken by computing as difference **VALID**(VPOS2)-**VALID**(VPOS1) the time elapsed between the passages of the vehicle at the two camera sites (assuming valid times have the granularity of 1 second, division by the constant 3,600 is needed to convert them to hours). If the 90 Km/h speed limit is exceeded, the query returns the identifier of the offending vehicle, the time of its passage at the C2 site and its average speed between C1 and C2.

Let us consider a vehicle making a passage at the C1 site at 12:00 a.m. recorded in the blockchain with a 15-minute delay, and a passage at the C2 site at 12:06 a.m. (of the same day) recorded in the blockchain with a 20-minute delay. Using BiTchain data, the query above using valid time is able to detect that the vehicle has covered the space between C1 and C2 in 6 minutes and, thus, travelled at a speed of 100 Km/h. With a traditional blockchain, without valid time, a query similar to the one in Listing 6 but using transaction time instead, would have evaluated the passage times of the vehicle at the C1 and C2 sites as 12:15 a.m. and 12:26 a.m., respectively. Hence, the elapsed time between the passages would have been computed as 11 minutes, corresponding to a transit average speed of 54.54 Km/h, seemingly under the speed limit. Therefore, with BiTchain data a fine shall be imposed on the vehicle, whereas the vehicle would get away with it (and the speed camera system becomes totally ineffective) if a traditional blockchain is used instead.

A more complex and powerful use of history variables is exemplified in the last query we present (in Listing 7). In such a case, two history variables V1 and V2 are declared to range over the transactions in the traffic blockchain, to denote the full history of the motion of two different vehicles. Then, two more variables, V1P1 and V1P2, are declared over V1 in order to denote two homogeneous versions within the V1 history characterized by a constant speed (the **PERIOD** keyword is used to define such versions by coalescing transactions of V1, with the same VX and VY values, that are consecutive along valid time). Similarly, a further history variable V2P1 is also declared over the history of the vehicle denoted by V2 in order to represent a homogeneous version with constant speed of such a vehicle.

*Listing 7. BiTEQL query example*

```
SELECT V2P1.VID, SPEED(V2P1)
FROM traffic.transactions AS VTR,
     VTR(VID) AS V1,V2,
     V1(VX,VY)(PERIOD) AS V1P1,V1P2,
     V2(VX,VY)(PERIOD) AS V2P1
WHERE V1P1 MEETS V1P2
  AND SPEED(V1P1) > 0 AND SPEED(V1P2) = 0
  AND BEGIN(V1P2) BETWEEN '2022-02-15 08:00'
                      AND '2022-02-15 08:30'
  AND DISTANCE(V1P2,(43.87,11.17)) < 10 METER
  AND V2P1 CONTAINS BEGIN(V1P2)
  AND DISTANCE(V2P1,V1P2) < 1 KILOMETER
ORDER BY V2P1.VID, VALID(V2P1);
```

The query in Listing 7 returns the identifier and speed of the vehicles V2 finding themselves within 1 Km of another vehicle V1, which stopped on the highway near (with a 10-meter approximation) the point with coordinates 43.87 N, 11.17 E, in the half hour after 8 a.m. on February 15, 2022. Vehicles V1 are determined using two consecutive versions of their position, V1P1 and V1P2, both with constant speed, such that they were moving in V1P1 (speed > 0) and sitting in V1P2 (speed = 0). In order to select vehicles V2, their version with constant speed V2P1 had to be less than 1 Km far from the position where V1 was sitting, at the time when V1 stopped. The results are ordered by vehicle identifier and valid time, so that it can be seen from consecutive listing of subsequent speeds if vehicles V2 were slowing down or not in approaching the sitting vehicle V1.

# RELATED WORK DISCUSSION

Temporal data models (Tansel, 1997; Wang & Zaniolo, 2004; Brahmia et al., 2016; Jensen & Snodgrass, 2018e) and temporal query languages (Snodgrass et al., 1995; Gao & Snodgrass, 2003; Grandi, 2010; Jensen & Snodgrass, 2018f) have been very well researched by the database community. In our present work, we propose a bitemporal data model and a temporal query language for blockchain to efficiently managing data of intelligent transportation and autonomous vehicles applications and more precisely those that require exact times of events.

## Temporal Data Models for Intelligent Transportation and Autonomous Vehicles

In the early 1990s, in the context of the Prometheus European project laying the foundations of modern intelligent transportation and autonomous vehicle technologies, Grandi et al. (1991) proposed to use a bitemporal database to keep track, for traffic control and accident investigation, of the information generated by vehicle-to-vehicle or vehicle-to-roadside communications.

Guo et al. (2018) propose a system, inspired from blockchain, to record accidents involving autonomous vehicles, for forensic purposes. More precisely, they introduce a mechanism of "Proof of Event with Dynamic Federation Consensus" to accurately record accident events as timestamped transactions in a new block, in real time.

Gupta et al. (2018a) deal with processing temporal queries on the blockchain platform Hyperledger Fabric. To overcome some problems related to temporal data querying in Fabric (data are organized in a file system, are accessed by end users through a limited API), they propose two models for creating temporal indexes on such a platform: the first one copies each event inserted by a Fabric transaction and stores temporally close events together; the second one does not make copies of events but tags some metadata to each event such that temporally close events share the same metadata. In (Gupta et al., 2018b), the authors have proposed variants to improve performances.

## Querying Spatio-temporal Blockchain

Qu et al. (2019) have proposed a blockchain that records time and location attributes for the transactions and supports spatial queries through the definition of a cryptographically signed tree, named the Merkle Block Space Index, which is an updated version of the Merkle KD-tree. The authors deal with temporal queries that work on a block-DAG structure, named Temporal Graph Search, without using temporal indexes.

Nasrulin et al. (2018) have introduced a proof-of-location spatio-temporal verification protocol for the blockchain. It supports an access control model and uses a set of verification rules to create and verify spatio-temporal data points. The authors show the usefulness of their protocol through the implementation of a system prototype (on top of the Hyperledger Iroha blockchain platform) in the supply chain context.

## Querying Traditional Blockchain

Li et al. (2017) proposed the EtherQL query language for the Ethereum blockchain platform. However, EtherQL queries are not executed on (block, account and transaction) data in Ethereum but on data in a MongoDB database that is synchronized with Ethereum. Notice that MongoDB is a famous NoSQL (Not Only SQL) document-oriented database management system. In (Pratama & Mutijarsa, 2018), the EtherQL language (Li et al., 2017) has been extended to support multiple search parameters and some analytic functions.

In (Bragagnolo et al., 2018), the Ethereum Query Language (EQL) has been proposed for Ethereum, as a SQL-like query language. To provide good performance when executing queries, EQL defines indexes on the hashes of blockchain data.

Lin and Sun (2018) propose a query language, named HyperQL, for the Hyperledger Fabric blockchain platform. HyperQL queries are not executed on data in Fabric but on data in a PostgreSQL database synchronized with Fabric. Notice that PostgreSQL is a very well-known NewSQL database management system and it has been chosen since it supports JOIN operations (which are not allowed by MongoDB).

In (Xu et al., 2019), the authors propose vChain, a framework that supports verifiable Boolean range queries over blockchain databases to guarantee the integrity of query results. The proposed framework is completed by some additional structures: (i) an authenticated data structure for dynamic aggregation over arbitrary query attributes, (ii) two indexing structures to aggregate intra-block and inter-block data records for efficient query verification, and (iii) a query index to speed up the processing of a large number of subscription queries simultaneously.

## Solutions for Inaccuracy of Block Timestamps in a Blockchain

In (Szalachowski, 2018), the author proposes a protocol for strengthening the reliability of Bitcoin timestamps. External timestamp authorities can be used to assert a block creation time, instead of only trusting timestamps put by miners. This extension can confirm that a block was created within a certain temporal interval. Moreover, it is low-cost since it does not need additional data structures nor changes to the Bitcoin protocol.

Zhang et al. (2019b) introduce Chronos, an accurate blockchain-based timestamping scheme, which deals with the single-point-of-failure problem. It supports an accurate way to prove that a file was created during a certain time interval. Chronos+ (Zhang et al., 2019a) extends this scheme by (i) supporting batch timestamping where multiple timestamping requests, from different users, can be satisfied simultaneously, (ii) securing the scheme from attacks of malicious competitors, and (iii) proposing the concept of WoT (window of timestamping) to measure the practicality of timestamping schemes. The experimental evaluation of the performance of Chronos+ shows that it is efficient w.r.t computation, communication, monetary costs and WoT. Some experiments have shown that Chronos+ could be also deployed for mobile users.

In the Bitcoin's Block Timestamp Protection Rules[7], published in 2019, the problem of malicious management of block timestamps in Bitcoin has been described (e.g., some nefarious miners provide future timestamps). Two rules have been proposed to protect the timestamps: (a) the Median Past Time Rule, where the timestamp must be further forwards than the median of the last eleven blocks, and (b) the Future Block Time Rule, where the timestamp cannot be more than 2 hours in the future, based on the MAX_FUTURE_BLOCK_TIME constant, relative to the median time from the node's peers. Recently, Ma et al. (2020) have also proposed a schema to achieve reliable blockchain timestamps for Bitcoin. The accuracy of block timestamps is narrowed to an average of ten minutes (instead of hours), by leveraging an outside trusted timestamp authority in the initial mining phase.

## Exploitation of Blockchain in Intelligent Transportation and Autonomous Vehicles Systems

Due to the rich literature on this topic and due to space limitations, we will restrict our survey to some representative papers that have been recently published.

Kim (2018) studied the key aspects of cybersecurity issues in intelligent connected vehicles and discussed methods for building blockchain-based secure trust networks among such intelligent vehicles.

Fu et al. (2020) considered this topic from the machine learning perspective and showed how blockchains could be applied to protect the distributed learned models.

From a security viewpoint, Diallo et al. (2020) proposed an architecture for secure data management in Vehicular Ad-hoc NETworks (VANETs) using blockchains.

From the angle of communication between vehicles, Rehman et al. (2020) proposed a blockchain-based vehicular network allowing a secure vehicle-to-vehicle communication.

Narbayeva et al. (2020) proposed to use blockchains for vehicle tracking and developed, on top of the fast Exonum blockchain platform, a safe and reliable system for tracking car actions. Parameters of the current state of each car are sent and confirmed through neighbors cars within a radius of 100-150 meters using a cryptographic scheme.

As to vehicle position correction, Li et al. (2020a) proposed a vehicular blockchain networks-based GPS error sharing framework that improves vehicle positioning accuracy by guaranteeing security and credibility of cooperators and data.

In (Abdel-Basset et al., 2021), the authors proposed a federated deep learning-based intrusion detection framework (FED-IDS), which allows to detect attacks by offloading the learning process from servers to distributed vehicular edge nodes. This framework introduces a context-aware transformer network to learn spatio-temporal representations of vehicular traffic flows necessary for classifying different categories of attacks. Blockchain-managed federated training is presented to enable multiple edge nodes to offer secure, distributed, and reliable training without the need for centralized authority.

Janakbhai et al. (2021) proposed to merge priority-based scheduling protocol with blockchain-based intelligent transportation system, and claimed that such a merging will reduce the network traffic and will not have any effect on the service ratio and reliability of transmission channel.

In (Ning et al., 2021), the authors constructed a blockchain-enabled crowdsensing framework for distributed traffic management in intelligent transportation systems. They define two algorithms for the functioning of such a framework: a Deep Reinforcement Learning (DRL)-based algorithm and a DIstributed Alternating Direction mEthod of Multipliers (DIADEM) algorithm. They also present experimental results that show that (i) the DRL-based algorithm can legitimately select active miners and transactions to make a satisfied trade-off between the blockchain safety and latency, and (ii) the DIADEM algorithm can effectively select task computation modes for vehicles in a distributed way to maximize their social welfare.

Li et al. (2021) first reviewed and compared the blockchain-based trust management approaches in cloud computing systems. Then, they proposed a novel cloud-edge hybrid trust management framework and a double-blockchain based cloud transaction model. Last, they gave some future research directions in blockchain-based trust management.

In (Zhao et al., 2022), the authors propose a blockchain-based effective renewable energy management process that allows to save renewable energies and to avoid wasteful usage. The blockchain process validates each electric vehicle (EV) before allowing it to be charged. The blockchain principle analyzes the energy demand of any EV and validates its demand through vehicle information. The validation process enables the same and original energy use vessel to be identified to efficiently remove threats associated to activities. The validating process considers the vehicle data and energy consumption level to check the EV. The block-based validation process monitors each EV entered into the grid environment, and the intelligent transportation system process minimizes the fuel wastage and enhances the system's overall efficiency.

Vishwakarma and Das (2022) proposed a new incentive mechanism for vehicles, named SmartCoin, based on a consortium blockchain. Such a mechanism is intended to provide the following advantages: social welfare, improved transportation, less traffic congestion, reduced road accidents, and a transportation network free from fraud messages. In this mechanism, vehicles assign the rating to the message source vehicle based on the validity of the message. Some SmartCoins are credited into the vehicle's account at the blockchain according to the rating of the message. The SmartCoins could be redeemed either at the fuel station, electric charge station for EVs and vehicle service station.

In (Das et al., 2022), the authors have dealt with the design and development of an intelligent transportation management system (ITMS) using blockchain and smart contracts. More precisely, a framework of blockchain-enabled automated toll-tax collection system (BATCS) is proposed as a blockchain-based ITMS application to collect the toll-tax amounts (TAs) without stopping vehicles while they pass the toll plaza. Smart contracts are used to authenticate vehicles' data and to collect TAs automatically. Moreover, the authors define an algorithm to efficiently check data and collect TAs. They state that the most important contributions of the BATCS with respect to the RFID-based system are less fuel consumption and time-saving for a vehicle. They also claim that their framework is secure, transparent, and privacy-preserving, in the world of the intelligent decentralized electronic toll collection (ETC) systems.

Tyagi et al. (2022) described the implementation of a system for securing future vehicles (and more precisely protecting personal information and habits of users of these vehicles) while using blockchain technology.

Modi et al. (2022) presented the software-defined networking (SDN) and cloud-enabled hierarchical future 5G-vehicular network architecture, its applications, issues, and challenges based on recent advances in technology and research.

In (Jabbar et al., 2022), the authors provide a systematic literature review of Blockchain technology for intelligent transportation systems, in general, and the IoV, in particular.

With regard to the state of the art, our contribution is mainly (i) the extension of the current blockchain, considered as of transaction time format, to the valid time, so that it becomes bitemporal and (ii) the proposal of an expressive temporal query language, designed as a TSQL2-like language, for extracting information from such a new blockchain. We think that this contribution is interesting and useful at both theoretical level, since it bridges the gap of the lack of a bitemporal data model and its associated temporal query language, in the blockchain world, and practical level, since it (a) provides a solution for applications that require accurate timestamps of events, in the field of intelligent transportation and autonomous vehicles, and (b) facilitates the exploitation of the huge amount of data stored in blockchain, through their querying via a user-friendly language.

## CONCLUSION

Intelligent transportation systems leverage blockchain technology thanks to its beneficial features like distribution, absence of central parties, neutral support of security and reliability of transactions. Nevertheless, some issues have not been efficiently handled so far, including those involving safety and those related to vehicle-to-vehicle communication and communication between vehicles and transport infrastructure components for refueling, charging, parking or repair services. We think that the resolution of these problematic issues can take advantage of temporal information that can be made available to the involved actors. However, standard blockchain timestamps often do not correspond with an acceptable accuracy to the occurrence time of events in the reality and, thus, are not useful for applications that require precise timing of events to be reconstructed from stored data like intelligent transportation and management of autonomous vehicles. Thus, in order to fulfill the requirements of such applications, we have presented in this chapter BiTchain, a bitemporal blockchain data model that extends the traditional blockchain data model with a new timestamp, which specifies the valid time of data that are provided by a data originator and written inside blockchain transactions. This time is assigned by the data producers to the data to be recorded, according to its local/network clock. In addition to BiTchain, we have also presented BiTEQL an expressive query language for BiTchain-based blockchains. This new language is a SQL-like query language equipped with both the basic constructs for querying traditional blockchains supported by EQL language (Bragagnolo et al., 2018) and the expressive temporal constructs which have been designed for the consensual temporal database query language TSQL2 (Snodgrass et al., 1995). The BiTchain data model and the BiTEQL query language have been first proposed in (Brahmia et al., 2022).

In our future work, we will design and implement a query engine that supports the BiTEQL language, possibly via the extension of a canonical EQL engine and the adoption of suitable index and storage structures to facilitate the execution of BiTEQL temporal queries on BiTchain-based blockchains.

## REFERENCES

Abdel-Basset, M., Moustafa, N., Hawash, H., Razzak, I., Sallam, K. M., & Elkomy, O. M. (2021). Federated Intrusion Detection in Blockchain-Based Smart Transportation Systems. *IEEE Transactions on Intelligent Transportation Systems*, *23*(3), 2523–2537. doi:10.1109/TITS.2021.3119968

Ali, M., Nelson, J., Shea, R., & Freedman, M. J. (2016). Blockstack: A global naming and storage system secured by blockchains. In *2016 USENIX annual technical conference (USENIX ATC'16)* (pp. 181-194). USENIX.

Allen, J. F. (1983). Maintaining knowledge about temporal intervals. *Communications of the ACM*, *26*(11), 832–843. doi:10.1145/182.358434

Anderson, J. M., Nidhi, K., Stanley, K. D., Sorensen, P., Samaras, C., & Oluwatola, O. A. (2014). *Autonomous vehicle technology: A guide for policymakers*. Rand Corporation.

Behnke, K., & Janssen, M. F. W. H. A. (2020). Boundary conditions for traceability in food supply chains using blockchain technology. *International Journal of Information Management*, *52*, 101969. doi:10.1016/j.ijinfomgt.2019.05.025

Bekmezci, I., Sahingoz, O. K., & Temel, Ş. (2013). Flying ad-hoc networks (FANETs): A survey. *Ad Hoc Networks*, *11*(3), 1254–1270. doi:10.1016/j.adhoc.2012.12.004

Bolour, A., Anderson, T. L., Dekeyser, L. J., & Wong, H. K. (1982). The role of time in information processing: A survey. *ACM SIGART Bulletin*, *80*(80), 28–46. doi:10.1145/1056176.1056180

Bragagnolo, S., Rocha, H., Denker, M., & Ducasse, S. (2018, May). Ethereum query language. In *Proceedings of the 1st International Workshop on Emerging Trends in Software Engineering for Blockchain* (pp. 1-8). Academic Press.

Brahmia, S., Brahmia, Z., Grandi, F., & Bouaziz, R. (2016, September). τJSchema: A Framework for Managing Temporal JSON-based NoSQL Databases. In *International Conference on Database and Expert Systems Applications (DEXA)* (pp. 167-181). Springer. 10.1007/978-3-319-44406-2_13

Brahmia, Z., Grandi, F. & Bouaziz, R. (2022). *A Bitemporal Blockchain Data Model and Query Language for Intelligent Transportation and Autonomous Vehicles Applications.* Submitted for Publication.

Chang, S. E., Luo, H. L., & Chen, Y. (2020). Blockchain-Enabled Trade Finance Innovation: A Potential Paradigm Shift on Using Letter of Credit. *Sustainability*, *12*(1), 188. doi:10.3390u12010188

Chondrogiannis, E., Andronikou, V., Karanastasis, E., Litke, A., & Varvarigou, T. (2022). Using blockchain and semantic web technologies for the implementation of smart contracts between individuals and health insurance organizations. *Blockchain: Research and Applications*, *3*(2), 100049. doi:10.1016/j.bcra.2021.100049

Das, D., Banerjee, S., Chatterjee, P., Biswas, M., Biswas, U., & Alnumay, W. (2022). Design and development of an intelligent transportation management system using blockchain and smart contracts. *Cluster Computing*, *25*(3), 1–15. doi:10.100710586-022-03536-z

Diallo, E. H., Al Agha, K., Dib, O., Laube, A., & Mohamed-Babou, H. (2020, April). Toward Scalable Blockchain for Data Management in VANETs. In *Workshops of the International Conference on Advanced Information Networking and Applications* (pp. 233-244). Springer. 10.1007/978-3-030-44038-1_22

Dinh, T. T. A., Liu, R., Zhang, M., Chen, G., Ooi, B. C., & Wang, J. (2018). Untangling blockchain: A data processing view of blockchain systems. *IEEE Transactions on Knowledge and Data Engineering*, *30*(7), 1366–1385. doi:10.1109/TKDE.2017.2781227

Frizzo-Barker, J., Chow-White, P. A., Adams, P. R., Mentanko, J., Ha, D., & Green, S. (2020). Blockchain as a disruptive technology for business: A systematic review. *International Journal of Information Management, 51*, 102029. doi:10.1016/j.ijinfomgt.2019.10.014

Fu, Y., Yu, F. R., Li, C., Luan, T. H., & Zhang, Y. (2020). Vehicular Blockchain-Based Collective Learning for Connected and Autonomous Vehicles. *IEEE Wireless Communications, 27*(2), 197–203. doi:10.1109/MNET.001.1900310

Gao, D., & Snodgrass, R. T. (2003, January). Temporal slicing in the evaluation of XML queries. In *Proceedings 2003 VLDB Conference* (pp. 632-643). Morgan Kaufmann. 10.1016/B978-012722442-8/50062-8

Grandi, F. (2010, September). T-SPARQL: A TSQL2-like Temporal Query Language for RDF. In ADBIS (Local Proceedings) (pp. 21-30). Academic Press.

Grandi, F. (2012). Introducing an annotated bibliography on temporal and evolution aspects in the semantic web. *SIGMOD Record, 41*(4), 18–21. doi:10.1145/2430456.2430460

Grandi, F. (2015). Temporal Databases. In M. Khosrow-Pour (Ed.), *Encyclopedia of Information Science and Technology* (3rd ed., pp. 1914–1922). IGI Global. doi:10.4018/978-1-4666-5888-2.ch184

Grandi, F., Scalas, M. R., & Tiberio, P. (1991). A Primitive for Recording the Content of PRO-COM Messages in Temporal Data Bases. *Proc. of 5th Prometheus Workshop*, 127–141.

Grandi, F., Scalas, M. R., & Tiberio, P. (1993). History and Tuple Variables for Temporal Query Languages. *Proc. of ARPA/NSF Workshop on an Infrastructure for Temporal Databases.*

Grandi, F., Scalas, M. R., & Tiberio, P. (1995). A History-oriented Temporal SQL Extension. In *1995 2nd International Workshop on Next Generation Information Technologies and Systems (NGITS)* (pp. 42-48). Technion Haifa.

Guo, H., Meamari, E., & Shen, C. C. (2018, August). Blockchain-inspired event recording system for autonomous vehicles. In *2018 1st IEEE International Conference on Hot Information-Centric Networking (HotICN)* (pp. 218-222). IEEE. 10.1109/HOTICN.2018.8606016

Gupta, H., Hans, S., Aggarwal, K., Mehta, S., Chatterjee, B., & Jayachandran, P. (2018a). Efficiently processing temporal queries on hyperledger fabric. In *2018 IEEE 34th International Conference on Data Engineering (ICDE)* (pp. 1489-1494). IEEE. 10.1109/ICDE.2018.00167

Gupta, H., Hans, S., Mehta, S., & Jayachandran, P. (2018b). On building efficient temporal indexes on hyperledger fabric. In *2018 IEEE 11th International Conference on Cloud Computing (CLOUD)* (pp. 294-301). IEEE. 10.1109/CLOUD.2018.00044

Gupta, R., Tanwar, S., Tyagi, S., & Kumar, N. (2020). Blockchain-based security attack resilience schemes for autonomous vehicles in industry 4.0: A systematic review. *Computers & Electrical Engineering*, *86*, 106717. doi:10.1016/j.compeleceng.2020.106717

Jabbar, R., Dhib, E., Said, A. B., Krichen, M., Fetais, N., Zaidan, E., & Barkaoui, K. (2022). Blockchain Technology for Intelligent Transportation Systems: A Systematic Literature Review. *IEEE Access: Practical Innovations, Open Solutions*, *10*, 20995–21031. doi:10.1109/ACCESS.2022.3149958

Janakbhai, N. D., Saurin, M. J., & Patel, M. (2021). Blockchain-Based Intelligent Transportation System with Priority Scheduling. In Data Science and Intelligent Applications. Lecture Notes on Data Engineering and Communications Technologies (vol. 52, pp. 311-317). Springer. doi:10.1007/978-981-15-4474-3_34

Javaid, M., Haleem, A., Singh, R. P., Khan, S., & Suman, R. (2021). Blockchain technology applications for Industry 4.0: A literature-based review. *Blockchain: Research and Applications*, *2*(4), 100027. doi:10.1016/j.bcra.2021.100027

Jensen, C. S., & Dyreson, C. E. (1998). The consensus glossary of temporal database concepts – February 1998 version. In Temporal Databases – Research and Practice (pp. 367–405). Berlin, Germany: Springer.

Jensen, C. S., & Snodgrass, R. T. (2018a). Temporal Database. In L. Liu & M. T. Özsu (Eds.), *Encyclopedia of Database Systems* (2nd ed.). Springer-Verlag. doi:10.1007/978-1-4614-8265-9_1415

Jensen, C. S., & Snodgrass, R. T. (2018b). Transaction Time. In L. Liu & M. T. Özsu (Eds.), *Encyclopedia of Database Systems*. Springer. doi:10.1007/978-1-4614-8265-9_1064

Jensen, C. S., & Snodgrass, R. T. (2018c). Valid Time. In L. Liu & M. T. Özsu (Eds.), *Encyclopedia of Database Systems*. Springer. doi:10.1007/978-1-4614-8265-9_1066

Jensen, C. S., & Snodgrass, R. T. (2018d). Bitemporal Relation. In L. Liu & M. T. Özsu (Eds.), *Encyclopedia of Database Systems*. Springer. doi:10.1007/978-1-4614-8265-9_1409

Jensen, C. S., & Snodgrass, R. T. (2018e). Temporal Data Models. In L. Liu & M. T. Özsu (Eds.), *Encyclopedia of Database Systems*. Springer. doi:10.1007/978-1-4614-8265-9_1415

Jensen, C. S., & Snodgrass, R. T. (2018f). Temporal Query Languages. In L. Liu & M. T. Özsu (Eds.), *Encyclopedia of Database Systems*. Springer. doi:10.1007/978-1-4614-8265-9_407

Khan, K. M., Arshad, J., & Khan, M. M. (2020). Investigating performance constraints for blockchain based secure e-voting system. *Future Generation Computer Systems*, *105*, 13–26. doi:10.1016/j.future.2019.11.005

Kher, R., Terjesen, S., & Liu, C. (2020). Blockchain, Bitcoin, and ICOs: A review and research agenda. *Small Business Economics*, *56*(4), 1699–1720. doi:10.100711187-019-00286-y

Kim, S. (2018). Blockchain for a trust network among intelligent vehicles. []. Elsevier.]. *Advances in Computers*, *111*, 43–68. doi:10.1016/bs.adcom.2018.03.010

Kulkarni, K., & Michels, J. E. (2012). Temporal features in SQL: 2011. *SIGMOD Record*, *41*(3), 34–43. doi:10.1145/2380776.2380786

Li, C., Fu, Y., Yu, F. R., Luan, T. H., & Zhang, Y. (2020a). Vehicle position correction: A vehicular blockchain networks-based GPS error sharing framework. *IEEE Transactions on Intelligent Transportation Systems*, *22*(2), 898–912. doi:10.1109/TITS.2019.2961400

Li, W., Wu, J., Cao, J., Chen, N., Zhang, Q., & Buyya, R. (2021). Blockchain-based trust management in cloud computing systems: A taxonomy, review and future directions. *Journal of Cloud Computing: Advances. Systems and Applications*, *10*(1), 1–34.

Li, X., Jiang, P., Chen, T., Luo, X., & Wen, Q. (2020b). A survey on the security of blockchain systems. *Future Generation Computer Systems*, *107*, 841–853. doi:10.1016/j.future.2017.08.020

Li, Y., Zheng, K., Yan, Y., Liu, Q., & Zhou, X. (2017, March). EtherQL: a query layer for blockchain system. In *International Conference on Database Systems for Advanced Applications (DASFAA)* (pp. 556-567). Springer. 10.1007/978-3-319-55699-4_34

Lin, Y. J., & Sun, M. T. (2018, October). HyperQL-Efficient Blockchain Query. In *Proc. of the 42th Taiwan Academic Network Conference (TANET)* (pp. 939-944). Academic Press.

Liu, J., Peng, S., Long, C., Wei, L., Liu, Y., & Tian, Z. (2020, March). Blockchain for Data Science. In *Proceedings of the 2020 The 2nd International Conference on Blockchain Technology* (pp. 24-28). 10.1145/3390566.3391681

Ma, G., Ge, C., & Zhou, L. (2020). Achieving reliable timestamp in the bitcoin platform. *Peer-to-Peer Networking and Applications, 13*(6), 2251–2259. doi:10.100712083-020-00905-6

Modi, Y., Panchal, M., Bhatia, J., & Tanwar, S. (2022). Blockchain-Based Software-Defined Vehicular Networks for Intelligent Transportation System Beyond 5G. In *Advances in Computing, Informatics, Networking and Cybersecurity* (pp. 513–533). Springer. doi:10.1007/978-3-030-87049-2_17

Mohan, C. (2018, April). Blockchains and databases: A new era in distributed computing. In *2018 IEEE 34th International Conference on Data Engineering (ICDE)* (pp. 1739-1740). IEEE. 10.1109/ICDE.2018.00227

Nakamoto, S. (2008). *Bitcoin: A peer-to-peer electronic cash system*. https://bitcoin.org/bitcoin.pdf

Narbayeva, S., Bakibayev, T., Abeshev, K., Makarova, I., Shubenkova, K., & Pashkevich, A. (2020). Blockchain Technology on the Way of Autonomous Vehicles Development. *Transportation Research Procedia, 44*, 168–175. doi:10.1016/j.trpro.2020.02.024

Nasrulin, B., Muzammal, M., & Qu, Q. (2018, November). A robust spatio-temporal verification protocol for blockchain. In *International Conference on Web Information Systems Engineering (WISE)* (pp. 52-67). Springer. 10.1007/978-3-030-02922-7_4

Ning, Z., Sun, S., Wang, X., Guo, L., Guo, S., Hu, X., Hu, B., & Kwok, R. (2021). Blockchain-enabled intelligent transportation systems: A distributed crowdsensing framework. *IEEE Transactions on Mobile Computing*, 1. Advance online publication. doi:10.1109/TMC.2021.3079984

Nofer, M., Gomber, P., Hinz, O., & Schiereck, D. (2017). Blockchain. *Business & Information Systems Engineering, 59*(3), 183–187. doi:10.100712599-017-0467-3

Pavithran, D., Shaalan, K., Al-Karaki, J. N., & Gawanmeh, A. (2020). Towards building a blockchain framework for IoT. *Cluster Computing, 23*(3), 2089–2103. doi:10.100710586-020-03059-5

Pratama, F. A., & Mutijarsa, K. (2018, October). Query Support for Data Processing and Analysis on Ethereum Blockchain. In *2018 International Symposium on Electronics and Smart Devices (ISESD)* (pp. 1-5). IEEE. 10.1109/ISESD.2018.8605476

Qu, Q., Nurgaliev, I., Muzammal, M., Jensen, C. S., & Fan, J. (2019). On spatio-temporal blockchain query processing. *Future Generation Computer Systems*, *98*, 208–218. doi:10.1016/j.future.2019.03.038

Rehman, M., Khan, Z. A., Javed, M. U., Iftikhar, M. Z., Majeed, U., Bux, I., & Javaid, N. (2020, April). A Blockchain Based Distributed Vehicular Network Architecture for Smart Cities. In *Workshops of the International Conference on Advanced Information Networking and Applications* (pp. 320-331). Springer. 10.1007/978-3-030-44038-1_29

Sharples, M., & Domingue, J. (2016, September). The blockchain and kudos: A distributed system for educational record, reputation and reward. In *European conference on technology enhanced learning* (pp. 490-496). Springer. 10.1007/978-3-319-45153-4_48

Singh, A., Parizi, R. M., Zhang, Q., Choo, K. K. R., & Dehghantanha, A. (2020). Blockchain smart contracts formalization: Approaches and challenges to address vulnerabilities. *Computers & Security*, *88*, 101654. doi:10.1016/j.cose.2019.101654

Six, N., Herbaut, N., & Salinesi, C. (2022). Blockchain software patterns for the design of decentralized applications: A systematic literature review. *Blockchain: Research and Applications*, *3*(2), 100061. doi:10.1016/j.bcra.2022.100061

Snodgrass. (Ed.). (1995). *The TSQL2 Temporal Query Language*. Kluwer Academic Publishers.

Szalachowski, P. (2018, June). (Short Paper) Towards More Reliable Bitcoin Timestamps. In *2018 Crypto Valley Conference on Blockchain Technology (CVCBT)* (pp. 101-104). IEEE. 10.1109/CVCBT.2018.00018

Taeihagh, A., & Lim, H. S. M. (2019). Governing autonomous vehicles: Emerging responses for safety, liability, privacy, cybersecurity, and industry risks. *Transport Reviews*, *39*(1), 103–128. doi:10.1080/01441647.2018.1494640

Tansel, A. U. (1997). Temporal relational data model. *IEEE Transactions on Knowledge and Data Engineering*, *9*(3), 464–479. doi:10.1109/69.599934

Tomescu, A., & Devadas, S. (2017, May). Catena: Efficient non-equivocation via bitcoin. In *2017 IEEE Symposium on Security and Privacy (SP)* (pp. 393-409). IEEE. 10.1109/SP.2017.19

Tschorsch, F., & Scheuermann, B. (2016). Bitcoin and beyond: A technical survey on decentralized digital currencies. *IEEE Communications Surveys and Tutorials*, *18*(3), 2084–2123. doi:10.1109/COMST.2016.2535718

Tyagi, A. K., Agarwal, D., & Sreenath, N. (2022, January). SecVT: Securing the Vehicles of Tomorrow using Blockchain Technology. In *2022 International Conference on Computer Communication and Informatics (ICCCI)* (pp. 1-6). IEEE. 10.1109/ICCCI54379.2022.9740965

Uddin, M. A., Stranieri, A., Gondal, I., & Balasubramanian, V. (2021). A survey on the adoption of blockchain in iot: Challenges and solutions. *Blockchain: Research and Applications*, *2*(2), 100006. doi:10.1016/j.bcra.2021.100006

Vishwakarma, L., & Das, D. (2022). SmartCoin: A novel incentive mechanism for vehicles in intelligent transportation system based on consortium blockchain. *Vehicular Communications*, *33*, 100429. doi:10.1016/j.vehcom.2021.100429

Vo, H. T., Kundu, A., & Mohania, M. K. (2018, March). Research Directions in Blockchain Data Management and Analytics. *Proceedings of EDBT*, *2018*, 445–448.

Wang, F., & Zaniolo, C. (2004, November). XBiT: an XML-based bitemporal data model. In *International Conference on Conceptual Modeling (ER)* (pp. 810-824). Springer.

Xu, C., Zhang, C., & Xu, J. (2019, June). vchain: Enabling verifiable boolean range queries over blockchain databases. In *Proceedings of the 2019 International Conference on Management of Data* (pp. 141-158). 10.1145/3299869.3300083

Yang, F., Wang, S., Li, J., Liu, Z., & Sun, Q. (2014). An overview of internet of vehicles. *China Communications*, *11*(10), 1–15. doi:10.1109/CC.2014.6969789

Zeadally, S., Hunt, R., Chen, Y. S., Irwin, A., & Hassan, A. (2012). Vehicular ad hoc networks (VANETS): Status, results, and challenges. *Telecommunication Systems*, *50*(4), 217–241. doi:10.100711235-010-9400-5

Zhang, P., & Boulos, M. N. K. (2020). Blockchain solutions for healthcare. In Precision Medicine for Investigators, Practitioners and Providers (pp. 519-524). Academic Press. doi:10.1016/B978-0-12-819178-1.00050-2

Zhang, Y., Xu, C., Cheng, N., Li, H., Yang, H., & Shen, X. (2019a). Chronos$^{+}$: An Accurate Blockchain-Based Time-Stamping Scheme for Cloud Storage. *IEEE Transactions on Services Computing*, *13*(2), 216–229. doi:10.1109/TSC.2019.2947476

Zhang, Y., Xu, C., Li, H., Yang, H., & Shen, X. S. (2019b). Chronos: Secure and accurate time-stamping scheme for digital files via blockchain. In *2019 IEEE International Conference on Communications (ICC)* (pp. 1-6). IEEE. 10.1109/ICC.2019.8762071

Zhao, J., He, C., Peng, C., & Zhang, X. (2022). Blockchain for effective renewable energy management in the intelligent transportation system. *Journal of Interconnection Networks, 2141009*. Advance online publication. doi:10.1142/S0219265921410097

Zheng, Z., Xie, S., Dai, H. N., Chen, X., & Wang, H. (2018). Blockchain challenges and opportunities: A survey. *International Journal of Web and Grid Services, 14*(4), 352–375. doi:10.1504/IJWGS.2018.095647

## KEY TERMS AND DEFINITIONS

**Autonomous Vehicle:** A vehicle that can, without interaction with a human driver, sense its environment and efficiently drive itself along several types of roads having different characteristics and possibly presenting complex and unexpected situations. It is also known as an automated or a self-driving vehicle, or as a driverless, self-driving or robotic car.

**Bitemporal Blockchain Data Model:** A blockchain data model that maintains the history of data along both transaction time (block timestamp, provided by the miner) and valid time (data validity timestamp, supplied by the data originator).

**Block Timestamp:** A component of a block, whose value is set by the miner to represent when the block has been validated. It gives to the blockchain the semantics of a transaction-time database.

**Blockchain:** A public distributed ledger that stores committed transactions in an ordered list, or a chain, of blocks. Such a ledger is maintained by multiple distributed nodes (computers) which are linked in a peer-to-peer network and possibly do not trust each other, through a consensus mechanism and cryptography. Notice that the ledger is replicated over all nodes.

**Intelligent Transportation System:** It is the application of sensing, analysis, control, and communications technologies to ground transportation in order to improve safety, mobility and efficiency

**Internet of Vehicles (IoV):** A wireless network of intelligent vehicles that are connected to Internet and that are communicating according to agreed protocols. IoV is a part of Internet of Things (IoT) and evolves from VANET.

**Temporal Data Model:** A data model for the representation of time-varying data.

**Temporal Database:** A database with built-in support for managing time-varying data.

**Temporal Query Language:** A query language that allows manipulation of time-referenced data.

**Transaction Time:** Temporal dimension concerning when some data is current in the database.

**TSQL2:** A temporal extension of the SQL standard designed by a committee of temporal database experts chaired by R.T. Snodgrass in 1995.

**Valid Time:** Temporal dimension concerning when some fact is true in the modeled reality.

**VANET:** A mobile network whose nodes are moving vehicles. In a VANET, intelligent vehicles on the road interact with each other and with roadside base stations.

## ENDNOTES

1 https://en.bitcoin.it/wiki/Block_timestamp

2 https://blog.bitmex.com/bitcoins-block-timestamp-protection-rules/

3 https://waymo.com/waymo-one/

4 https://www.theverge.com/2020/3/10/21172213/uber-self-driving-car-resume-testing-san-francisco-crash

5 https://www.tesla.com/support/autopilot

6 https://www.tesla.com/support/autopilot

7 https://blog.bitmex.com/bitcoins-block-timestamp-protection-rules/

# Compilation of References


Abdel-Basset, M., Moustafa, N., Hawash, H., Razzak, I., Sallam, K. M., & Elkomy, O. M. (2021). Federated Intrusion Detection in Blockchain-Based Smart Transportation Systems. *IEEE Transactions on Intelligent Transportation Systems*, *23*(3), 2523–2537. doi:10.1109/TITS.2021.3119968

Abuashour, A., & Kadoch, M. (2017). Performance improvement of cluster-based routing protocol in VANET. *IEEE Access: Practical Innovations, Open Solutions*, *5*, 15354–15371. doi:10.1109/ACCESS.2017.2733380

Adão, T., Jonáš, H., Luís, P., José, B., Emanuel, P., & Raul, M. (2017). Hyperspectral imaging: A review on UAV-based sensors, data processing and applications for agriculture and forestry. *Remote Sensing*, *9*(11), 1110. doi:10.3390/rs9111110

Agrawal, P., & Vutukuru, M. (2016). Trace based application layer modeling in ns-3. *2016 Twenty Second National Conference on Communication (NCC)*, 1-6. 10.1109/NCC.2016.7561126

Ahmad, F., Franqueira, V. N., & Adnane, A. (2018). TEAM: A trust evaluation and management framework in context-enabled vehicular ad-hoc networks. *IEEE Access: Practical Innovations, Open Solutions*, *6*, 28643–28660. doi:10.1109/ACCESS.2018.2837887

Ahmed, H., Pierre, S., & Quintero, A. (2017). A flexible testbed architecture for VANET. *Vehicular Communications*, 115-126.

Ahmed, G. A., Sheltami, T. R., Mahmoud, A. S., Imran, M., & Shoaib, M. (2021). A novel collaborative IoD-assisted VANET approach for coverage area maximization. *IEEE Access: Practical Innovations, Open Solutions*, *9*, 61211–61223. doi:10.1109/ACCESS.2021.3072431

Akhtar, N., Ergen, S. C., & Ozkasap, O. (2014). Vehicle mobility and communication channel models for realistic and efficient highway VANET simulation. *IEEE Transactions on Vehicular Technology*, *64*(1), 248–262. doi:10.1109/TVT.2014.2319107

Al-Absi, M. A.-A., & Lee, H. (2021). Moving ad hoc networks—A comparative study. *Sustainability*, 61–87.

Alghamdi, Y., Munir, A., & La, H. M. (2021). Architecture, Classification, and Applications of Contemporary Unmanned Aerial Vehicles. *IEEE Consumer Electronics Magazine*, 9--20.

Alharthi, A., Ni, Q., & Jiang, R. (2021). A privacy-preservation framework based on biometrics blockchain (BBC) to prevent attacks in VANET. *IEEE Access: Practical Innovations, Open Solutions*, *9*, 87299–87309. doi:10.1109/ACCESS.2021.3086225

Ali, M., Nelson, J., Shea, R., & Freedman, M. J. (2016). Blockstack: A global naming and storage system secured by blockchains. In *2016 USENIX annual technical conference (USENIX ATC'16)* (pp. 181-194). USENIX.

Allen, J. F. (1983). Maintaining knowledge about temporal intervals. *Communications of the ACM*, *26*(11), 832–843. doi:10.1145/182.358434

Almagbile, A. (2019). Estimation of crowd density from UAVs images based on corner detection procedures and clustering analysis. *Geo-Spatial Information Science*, *22*(1), 23–34. doi:10.1080/10095020.2018.1539553

Alotaibi, E. T., Alqefari, S. S., & Koubaa, A. (2019). Lsar: Multi-uav collaboration for search and rescue missions. *IEEE Access: Practical Innovations, Open Solutions*, *7*, 55817–55832. doi:10.1109/ACCESS.2019.2912306

Al-Roubaiey, A., & Al-Jamimi, H. (2019, June). Online power Tossim simulator for wireless sensor networks. *2019 11th International Conference on Electronics, Computers and Artificial Intelligence (ECAI)*, 1-5. 10.1109/ECAI46879.2019.9042005

Alshbatat, A. I., & Dong, L. (2010). Adaptive MAC Protocol for UAV Communication Networks Using Directional Antennas. *2010 International Conference on Networking, Sensing and Control (ICNSC)*, 598–603. 10.1109/ICNSC.2010.5461589

Al-Sheary, A., & Almagbile, A. (2017). Crowd monitoring system using unmanned aerial vehicle (UAV). *Journal of Civil Engineering and Architecture*, *11*(11), 1014–1024. doi:10.17265/1934-7359/2017.11.004

Altawy, R., & Youssef, A. M. (2016). Security, privacy, and safety aspects of civilian drones: A survey. *ACM Transactions on Cyber-Physical Systems*, *1*(2), 1–25. doi:10.1145/3001836

Alvear, O., Zema, N. R., Natalizio, E., & Calafate, C. T. (2017). Using UAV-based systems to monitor air pollution in areas with poor accessibility. *Journal of Advanced Transportation*, *2017*. doi:10.1155/2017/8204353

Amadeo, M., Campolo, C., & Molinaro, A. (2012). Enhancing IEEE 802.11 p/WAVE to provide infotainment applications in VANETs. *Ad Hoc Networks*, *10*(2), 253–269. doi:10.1016/j.adhoc.2010.09.013

Ambrosia, V., & Zajkowski, T. (2015). Selection of appropriate class UAS/sensors to support fire monitoring: experiences in the United States. Handbook of Unmanned Aerial Vehicles, 2723-2754.

Amoozadeh, M., Ching, B., Chuah, C. N., Ghosal, D., & Zhang, H. M. (2019). VENTOS: Vehicular network open simulator with hardware-in-the-loop support. *Procedia Computer Science*, *151*, 61–68. doi:10.1016/j.procs.2019.04.012

Anderson, J. M., Nidhi, K., Stanley, K. D., Sorensen, P., Samaras, C., & Oluwatola, O. A. (2014). *Autonomous vehicle technology: A guide for policymakers*. Rand Corporation.

Anjum, S. S., Noor, R., Ahmedy, I., Anisi, M. H., Azzuhri, S. R., Kiah, M. L. M., ... Kumar, P. (2020). An Optimal Management Modelling of Energy Harvesting and Transfer for IoT-based RF-enabled Sensor Networks. *Ad-Hoc & Sensor Wireless Networks*, 46.

Arafat, M. Y., & Moh, S. (2018). A survey on cluster-based routing protocols for unmanned aerial vehicle networks. *IEEE Access: Practical Innovations, Open Solutions*, *7*, 498–516. doi:10.1109/ACCESS.2018.2885539

Arena, F., & Pau, G. (2019). *An overview of vehicular communications*. Future Internet. doi:10.3390/fi11020027

Arnosti, S. Z., Pires, R. M., & Branco, K. R. (2017, June). Evaluation of cryptography applied to broadcast storm mitigation algorithms in FANETs. In *2017 International Conference on Unmanned Aircraft Systems (ICUAS)* (pp. 1368-1377). IEEE. 10.1109/ICUAS.2017.7991377

Asadpour, M., Giustiniano, D., Hummel, K. A., & Heimlicher, S. (2013, July). Characterizing 802.11 n aerial communication. In *Proceedings of the second ACM MobiHoc workshop on Airborne networks and communications* (pp. 7-12). 10.1145/2491260.2491262

Aznar-Poveda, J., Garcia-Sanchez, A. J., Egea-Lopez, E., & Garcia-Haro, J. (2021). Mdprp: A q-learning approach for the joint control of beaconing rate and transmission power in vanets. *IEEE Access: Practical Innovations, Open Solutions*, *9*, 10166–10178. doi:10.1109/ACCESS.2021.3050625

Bacco, M., Cassará, P., Colucci, M., Gotta, A., Marchese, M., & Patrone, F. (2017, September). A survey on network architectures and applications for nanosat and UAV swarms. In *International Conference on Wireless and Satellite Systems* (pp. 75-85). Springer.

Behnke, K., & Janssen, M. F. W. H. A. (2020). Boundary conditions for traceability in food supply chains using blockchain technology. *International Journal of Information Management*, *52*, 101969. doi:10.1016/j.ijinfomgt.2019.05.025

Bejiga, M. B., Zeggada, A., Nouffidj, A., & Melgani, F. (2016). A convolutional neural network approach for assisting avalanche search and rescue operations with UAV imagery. *Remote Sensing*, 100.

Bekmezci, I., Sahingoz, O. K., & Temel, Ş. (2013). Flying ad-hoc networks (FANETs): A survey. *Ad Hoc Networks*, *11*(3), 1254–1270. doi:10.1016/j.adhoc.2012.12.004

Belbachir, A., Escareno, J., Rubio, E., & Sossa, H. (2015). Preliminary results on UAV-based forest fire localization based on decisional navigation. In *Workshop on Research, Education and Development of Unmanned Aerial Systems (RED-UAS)* (pp. 377--382). IEEE. 10.1109/RED-UAS.2015.7441030

Berni, J. A., Zarco-Tejada, P. J., Suarez, L., & Fereres, E. (2009). Thermal and narrowband multispectral remote sensing for vegetation monitoring from an unmanned aerial vehicle. *IEEE Transactions on Geoscience and Remote Sensing*, *47*(3), 722–738. doi:10.1109/TGRS.2008.2010457

Bhoi, S. K., Puthal, D., Khilar, P. M., Rodrigues, J. J., Panda, S. K., & Yang, L. T. (2018). Adaptive routing protocol for urban vehicular networks to support sellers and buyers on wheels. *Computer Networks*, *142*, 168–178. doi:10.1016/j.comnet.2018.05.024

Bittar, A., & de Oliveira, N. M. (2013). Central processing unit for an autopilot: Description and hardware-in-the-loop simulation. *Journal of Intelligent & Robotic Systems*, *70*(1), 557–574. doi:10.100710846-012-9745-y

Bolour, A., Anderson, T. L., Dekeyser, L. J., & Wong, H. K. (1982). The role of time in information processing: A survey. *ACM SIGART Bulletin*, *80*(80), 28–46. doi:10.1145/1056176.1056180

Bragagnolo, S., Rocha, H., Denker, M., & Ducasse, S. (2018, May). Ethereum query language. In *Proceedings of the 1st International Workshop on Emerging Trends in Software Engineering for Blockchain* (pp. 1-8). Academic Press.

Brahmia, Z., Grandi, F. & Bouaziz, R. (2022). *A Bitemporal Blockchain Data Model and Query Language for Intelligent Transportation and Autonomous Vehicles Applications.* Submitted for Publication.

Brahmia, S., Brahmia, Z., Grandi, F., & Bouaziz, R. (2016, September). τJSchema: A Framework for Managing Temporal JSON-based NoSQL Databases. In *International Conference on Database and Expert Systems Applications (DEXA)* (pp. 167-181). Springer. 10.1007/978-3-319-44406-2_13

Buchenscheit, A., Schaub, F., Kargl, F., & Weber, M. (2009). A VANET-based emergency vehicle warning system. *IEEE Vehicular Networking Conference (VNC)* (p. 2009). IEEE.

Bujari, A., Palazzi, C. E., & Ronzani, D. 2017, June. FANET application scenarios and mobility models. In *Proceedings of the 3rd Workshop on Micro Aerial Vehicle Networks, Systems, and Applications* (pp. 43-46). 10.1145/3086439.3086440

Byun, S., Shin, I.-K., Moon, J., Kang, J., & Choi, S.-I. (2021). Road traffic monitoring from UAV images using deep learning networks. *Remote Sensing*, *13*(20), 4027. doi:10.3390/rs13204027

Cai, Y., Yu, F. R. R., Li, J., Zhou, Y., & Lamont, L. (2012). *MAC Performance Improvement in UAV Ad-Hoc Networks with Full-Duplex Radios and Multi-Packet Reception Capability.* Advance online publication. doi:10.1109/ICC.2012.6364116

Campion, M., Ranganathan, P., & Faruque, S. (2018, May). A review and future directions of UAV swarm communication architectures. In *2018 IEEE international conference on electro/information technology (EIT)* (pp. 903-908). IEEE.

Cao, S., & Lee, V. C. (2020). An accurate and complete performance modeling of the IEEE 802.11 p MAC sublayer for VANET. *Computer Communications*, *149*, 107–120. doi:10.1016/j.comcom.2019.08.026

Chakraborty, A., Chai, E., Sundaresan, K., Khojastepour, A., & Rangarajan, S. (2018, December). SkyRAN: a self-organizing LTE RAN in the sky. In *Proceedings of the 14th International Conference on emerging Networking EXperiments and Technologies* (pp. 280-292). 10.1145/3281411.3281437

Chang, S. E., Luo, H. L., & Chen, Y. (2020). Blockchain-Enabled Trade Finance Innovation: A Potential Paradigm Shift on Using Letter of Credit. *Sustainability*, *12*(1), 188. doi:10.3390u12010188

Chaurasia, B. K., & Verma, S. (2014). Secure pay while on move toll collection using VANET. *Computer Standards & Interfaces*, 403-411.

Cheng, C. M., & Tsao, S. L. (2014). Adaptive lookup protocol for two-tier VANET/P2P information retrieval services. *IEEE Transactions on Vehicular Technology*, *64*(3), 1051–1064. doi:10.1109/TVT.2014.2329015

Chen, J., Zhou, H., Zhang, N., Xu, W., Yu, Q., Gui, L., & Shen, X. (2017). Service-oriented dynamic connection management for software-defined internet of vehicles. *IEEE Transactions on Intelligent Transportation Systems*, *18*(10), 2826–2837. doi:10.1109/TITS.2017.2705978

Chondrogiannis, E., Andronikou, V., Karanastasis, E., Litke, A., & Varvarigou, T. (2022). Using blockchain and semantic web technologies for the implementation of smart contracts between individuals and health insurance organizations. *Blockchain: Research and Applications*, *3*(2), 100049. doi:10.1016/j.bcra.2021.100049

Chriki, A., Touati, H., Snoussi, H., & Kamoun, F. (2019, June). UAV-GCS centralized data-oriented communication architecture for crowd surveillance applications. In *2019 15th International Wireless Communications & Mobile Computing Conference (IWCMC)* (pp. 2064-2069). IEEE.

Chriki, A., Touati, H., Snoussi, H., & Kamoun, F. (2019). FANET: Communication, mobility models and security issues. *Computer Networks*, *163*, 106877. doi:10.1016/j.comnet.2019.106877

Contreras-Castillo, J., Zeadally, S., & Guerrero Ibáñez, J. A. (2017a). A seven-layered model architecture for internet of vehicles. *Journal of Information and Telecommunication*, *1*(1), 4–22. doi:10.1080/24751839.2017.1295601

Contreras-Castillo, J., Zeadally, S., & Guerrero-Ibañez, J. A. (2017b). Internet of vehicles: Architecture, protocols, and security. *IEEE Internet of Things Journal*, *5*(5), 3701–3709. doi:10.1109/JIOT.2017.2690902

Contreras, M., & Gamess, E. (2020). Real-Time Counting of Vehicles Stopped at a Traffic Light Using Vehicular Network Technology. *IEEE Access: Practical Innovations, Open Solutions*, *8*, 135244–135263. doi:10.1109/ACCESS.2020.3011195

Corridor simulation (CORSIM/TSIS). (n.d.). *Traffic Analysis Tools: Corridor Simulation - FHWA Operations*. https://ops.fhwa.dot.gov/trafficanalysistools/corsim.htm

Cumino, P., Lobato, W. Junior, Tavares, T., Santos, H., Rosário, D., Cerqueira, E., ... Gerla, M. (2018). Cooperative UAV scheme for enhancing video transmission and global network energy efficiency. *Sensors (Basel)*, *18*(12), 4155. doi:10.339018124155 PMID:30486376 da Cruz, E. P. F. (2018). A comprehensive survey in towards to future FANETs. *IEEE Latin America Transactions, 16*(3), 876–884. doi:10.1109/TLA.2018.8358668

Dai, R., Fotedar, S., Radmanesh, M., & Kumar, M. (2018). Quality-aware UAV coverage and path planning in geometrically complex environments. *Ad Hoc Networks, 73*, 95–105. doi:10.1016/j.adhoc.2018.02.008

Das, D., Banerjee, S., Chatterjee, P., Biswas, M., Biswas, U., & Alnumay, W. (2022). Design and development of an intelligent transportation management system using blockchain and smart contracts. *Cluster Computing, 25*(3), 1–15. doi:10.100710586-022-03536-z

Deeksha, M., Patil, A., Kulkarni, M., Shet, N. S. V., & Muthuchidambaranathan, P. (2022). Multistate Active Combined Power and Message/Data Rate Adaptive Decentralized Congestion Control Mechanisms for Vehicular Ad Hoc Networks. *Journal of Physics: Conference Series, 161*(1), 012018. doi:10.1088/1742-6596/2161/1/012018

Diallo, E. H., Al Agha, K., Dib, O., Laube, A., & Mohamed-Babou, H. (2020, April). Toward Scalable Blockchain for Data Management in VANETs. In *Workshops of the International Conference on Advanced Information Networking and Applications* (pp. 233-244). Springer. 10.1007/978-3-030-44038-1_22

Ding , Q. ( 2019 ). A mathematical model for reflection of electromagnetic wave. *Journal of Physics: Conference Series, 1213*(4).

Dinh, T. T. A., Liu, R., Zhang, M., Chen, G., Ooi, B. C., & Wang, J. (2018). Untangling blockchain: A data processing view of blockchain systems. *IEEE Transactions on Knowledge and Data Engineering, 30*(7), 1366–1385. doi:10.1109/TKDE.2017.2781227

Din, I. U., Ahmad, B., Almogren, A., Almajed, H., Mohiuddin, I., & Rodrigues, J. J. (2020). Left-right-front caching strategy for vehicular networks in icn-based internet of things. *IEEE Access: Practical Innovations, Open Solutions, 9*, 595–605. doi:10.1109/ACCESS.2020.3046887

Fadlullah, Z. M., Takaishi, D., Nishiyama, H., Kato, N., & Miura, R. (2016). A dynamic trajectory control algorithm for improving the communication throughput and delay in UAV-aided networks. *IEEE Network, 30*(1), 100–105. doi:10.1109/MNET.2016.7389838

Fan, X., Cai, W., & Lin, J. (2017, October). A survey of routing protocols for highly dynamic mobile ad hoc networks. In *2017 IEEE 17th International Conference on Communication Technology (ICCT)* (pp. 1412-1417). IEEE. 10.1109/ICCT.2017.8359865

Fandetti, G. N. (2017). *Method of drone delivery using aircraft.* U.S. Patent Application 14/817,356.

Fatemidokht, H., Rafsanjani, M. K., Gupta, B. B., & Hsu, C.-H. (2021). Efficient and secure routing protocol based on artificial intelligence algorithms with UAV-assisted for vehicular ad hoc networks in intelligent transportation systems. *IEEE Transactions on Intelligent Transportation Systems, 22*(7), 4757–4769. doi:10.1109/TITS.2020.3041746

Feiri, M., Petit, J., Schmidt, R. K., & Kargl, F. (2013). The impact of security on cooperative awareness in VANET. *Vehicular Networking Conference (VNC)*, 127-134. 10.1109/VNC.2013.6737599

Feng, K., Li, W., Ge, S., & Pan, F. (2020). Packages delivery based on marker detection for UAVs. In *Chinese Control and Decision Conference (CCDC)* (pp. 2094--2099). IEEE. 10.1109/CCDC49329.2020.9164677

Fitah, A., Badri, A., Moughit, M., & Sahe, A. (2018). Performance of DSRC and WIFI for Intelligent Transport Systems in VANET. *Procedia Computer Science*, *127*, 360–368. doi:10.1016/j.procs.2018.01.133

Frew, E. W., & Brown, T. X. (2008). Networking issues for small unmanned aircraft systems. *Journal of Intelligent & Robotic Systems*, *54*(1), 21–37.

Friis, H. T. (1946). A note on a simple transmission formula. *Proceedings of the IRE*, 254-256. 10.1109/JRPROC.1946.234568

Frizzo-Barker, J., Chow-White, P. A., Adams, P. R., Mentanko, J., Ha, D., & Green, S. (2020). Blockchain as a disruptive technology for business: A systematic review. *International Journal of Information Management*, *51*, 102029. doi:10.1016/j.ijinfomgt.2019.10.014

Fu, Y., Yu, F. R., Li, C., Luan, T. H., & Zhang, Y. (2020). Vehicular Blockchain-Based Collective Learning for Connected and Autonomous Vehicles. *IEEE Wireless Communications*, *27*(2), 197–203. doi:10.1109/MNET.001.1900310

Gao, D., & Snodgrass, R. T. (2003, January). Temporal slicing in the evaluation of XML queries. In *Proceedings 2003 VLDB Conference* (pp. 632-643). Morgan Kaufmann. 10.1016/B978-012722442-8/50062-8

Garcia-Ruiz, F., Sankaran, S., Maja, J. M., Lee, W. S., Rasmussen, J., & Ehsani, R. (2013). Comparison of two aerial imaging platforms for identification of Huanglongbing-infected citrus trees. *Computers and Electronics in Agriculture*, *91*, 106–115. doi:10.1016/j.compag.2012.12.002

Geipel, J., Link, J., & Claupein, W. (2014). Combined spectral and spatial modeling of corn yield based on aerial images and crop surface models acquired with an unmanned aircraft system. *Remote Sensing*, *6*(11), 10335–10355. doi:10.3390/rs61110335

Gheisari, M., Irizarry, J., & Walker, B. N. (2014). UAS4SAFETY: The potential of unmanned aerial systems for construction safety applications. In *Construction Research Congress 2014: Construction in a Global Network* (pp. 1801-1810). Academic Press.

Glass, S., Mahgoub, I., & Rathod, M. (2017). Leveraging manet-based cooperative cache discovery techniques in vanets: A survey and analysis. *IEEE Communications Surveys and Tutorials*, *19*(4), 2640–2661. doi:10.1109/COMST.2017.2707926

Gong, J., Chang, T. H., Shen, C., & Chen, X. (2018). Flight time minimization of UAV for data collection over wireless sensor networks. *IEEE Journal on Selected Areas in Communications*, *36*(9), 1942–1954. doi:10.1109/JSAC.2018.2864420

Gonzalez-Dugo, V., Zarco-Tejada, P., Nicolás, E., Nortes, P. A., Alarcón, J. J., Intrigliolo, D. S., & Fereres, E. J. P. A. (2013). Using high resolution UAV thermal imagery to assess the variability in the water status of five fruit tree species within a commercial orchard. *Precision Agriculture, 14*(6), 660–678. doi:10.100711119-013-9322-9

Goyal, A. K., Agarwal, G., & Tripathi, A. K. (2019). Network Architectures, Challenges, Security Attacks, Research Domains and Research Methodologies in VANET: A Survey. *International Journal of Computer Network & Information Security*, 37-44.

Gozalvez, J., Sepulcre, M., & Bauza, R. (2012). Impact of the radio channel modelling on the performance of VANET communication protocols. *Telecommunication Systems, 50*(3), 149–167. doi:10.100711235-010-9396-x

Grandi, F. (2010, September). T-SPARQL: A TSQL2-like Temporal Query Language for RDF. In ADBIS (Local Proceedings) (pp. 21-30). Academic Press.

Grandi, F., Scalas, M. R., & Tiberio, P. (1995). A History-oriented Temporal SQL Extension. In *1995 2nd International Workshop on Next Generation Information Technologies and Systems (NGITS)* (pp. 42-48). Technion Haifa.

Grandi, F. (2012). Introducing an annotated bibliography on temporal and evolution aspects in the semantic web. *SIGMOD Record, 41*(4), 18–21. doi:10.1145/2430456.2430460

Grandi, F. (2015). Temporal Databases. In M. Khosrow-Pour (Ed.), *Encyclopedia of Information Science and Technology* (3rd ed., pp. 1914–1922). IGI Global. doi:10.4018/978-1-4666-5888-2.ch184

Grandi, F., Scalas, M. R., & Tiberio, P. (1991). A Primitive for Recording the Content of PRO-COM Messages in Temporal Data Bases. *Proc. of 5th Prometheus Workshop*, 127–141.

Grandi, F., Scalas, M. R., & Tiberio, P. (1993). History and Tuple Variables for Temporal Query Languages. *Proc. of ARPA/NSF Workshop on an Infrastructure for Temporal Databases.*

Grzybowski, J., Latos, K., & Czyba, R. (2020). *Low-cost autonomous UAV-based solutions to package delivery logistics*. Springer. doi:10.1007/978-3-030-50936-1_42

Guo, H., Meamari, E., & Shen, C. C. (2018, August). Blockchain-inspired event recording system for autonomous vehicles. In *2018 1st IEEE International Conference on Hot Information-Centric Networking (HotICN)* (pp. 218-222). IEEE. 10.1109/HOTICN.2018.8606016

Guo, L., Dong, M., Ota, K., Li, Q., Ye, T., Wu, J., & Li, J. (2017, April). A secure mechanism for big data collect in in large scale internet of vehicle. *IEEE Internet of Things Journal, 4*(2), 601610.

Gupta, H., Hans, S., Aggarwal, K., Mehta, S., Chatterjee, B., & Jayachandran, P. (2018a). Efficiently processing temporal queries on hyperledger fabric. In *2018 IEEE 34th International Conference on Data Engineering (ICDE)* (pp. 1489-1494). IEEE. 10.1109/ICDE.2018.00167

Gupta, H., Hans, S., Mehta, S., & Jayachandran, P. (2018b). On building efficient temporal indexes on hyperledger fabric. In *2018 IEEE 11th International Conference on Cloud Computing (CLOUD)* (pp. 294-301). IEEE. 10.1109/CLOUD.2018.00044

Gupta, R., Tanwar, S., Tyagi, S., & Kumar, N. (2020). Blockchain-based security attack resilience schemes for autonomous vehicles in industry 4.0: A systematic review. *Computers & Electrical Engineering*, *86*, 106717. doi:10.1016/j.compeleceng.2020.106717

Haghighi, M. S., & Aziminejad, Z. (2019). Highly anonymous mobility-tolerant location-based onion routing for VANETs. *IEEE Internet of Things Journal*, *7*(4), 2582–2590. doi:10.1109/JIOT.2019.2948315

Hamdi, M. M., Audah, L., & Rashid, S. A. (2022). Data dissemination in VANETs using clustering and probabilistic forwarding based on adaptive jumping multi-objective firefly optimization. *IEEE Access: Practical Innovations, Open Solutions*, *10*, 14624–14642. doi:10.1109/ACCESS.2022.3147498

Hamdi, M. M., Audah, L., Rashid, S. A., Mohammed, A. H., Alani, S., & Mustafa, A. S. (2020). A review of applications, characteristics and challenges in vehicular ad hoc networks (VANETs). In *International Congress on Human-Computer Interaction, Optimization and Robotic Applications (HORA)* (pp. 1-7). IEEE.

Hartenstein, H., & Laberteaux, K. P. (2008). A tutorial survey on vehicular Ad Hoc Networks. *IEEE Communications Magazine*, *46*(June), 164–171. doi:10.1109/MCOM.2008.4539481

Hassanalian, M., & Abdelkefi, A. (2017). Classifications, applications, and design challenges of drones: A review. *Progress in Aerospace Sciences*, *91*, 99–131. doi:10.1016/j.paerosci.2017.04.003

Hayat, S., Yanmaz, E., Brown, T. X., & Christian, C. (2017). Multi-objective UAV path planning for search and rescue. In IEEE international conference on robotics and automation (ICRA) (pp. 5569-5574). IEEE. doi:10.1109/ICRA.2017.7989656

Hayat, S., Yanmaz, E., & Muzaffar, R. (2016). Survey on unmanned aerial vehicle networks for civil applications: A communications viewpoint. *IEEE Communications Surveys and Tutorials*, *18*(4), 2624–2661. doi:10.1109/COMST.2016.2560343

Helen, D., & Arivazhagan, D. (2014). Applications, advantages and challenges of ad hoc networks. *Journal of Academia and Industrial Research*, 453–457.

Hinzmann, T., Stastny, T., Conte, G., Doherty, P., Rudol, P., Wzorek, M., ... Gilitschenski, I. (2016, October). Collaborative 3d reconstruction using heterogeneous uavs: System and experiments. In *International Symposium on Experimental Robotics* (pp. 43-56). Springer.

Hong, Y., Fang, J., & Tao, Y. (2008, October). Ground control station development for autonomous UAV. In *International Conference on Intelligent Robotics and Applications* (pp. 36-44). Springer. 10.1007/978-3-540-88518-4_5

Huang, H., Savkin, A. V., & Huang, C. (2020). Scheduling of a parcel delivery system consisting of an aerial drone interacting with public transportation vehicles. *Sensors (Basel), 20*(7), 20–45. doi:10.339020072045 PMID:32260583

Huang, H., Savkin, A. V., & Huang, C. (2021). Decentralized autonomous navigation of a UAV network for road traffic monitoring. *IEEE Transactions on Aerospace and Electronic Systems, 57*(4), 2558–2564. doi:10.1109/TAES.2021.3053115

Huang, W., Li, P., & Zhang, T. (2018). RSUs placement based on vehicular social mobility in VANETs. In *IEEE Conference on Industrial Electronics and Applications (ICIEA)* (pp. 1255-1260). IEEE. 10.1109/ICIEA.2018.8397902

Huang, Y., Thomson, S. J., Hoffmann, W. C., Lan, Y., & Fritz, B. K. (2013). Development and prospect of unmanned aerial vehicle technologies for agricultural production management. *International Journal of Agricultural and Biological Engineering, 6*(3), 1–10.

Hu, X., Wu, T., & Wang, Y. (2019). Social coalition-based v2v broadcasting optimization algorithm in vanets. In *International Conference on Swarm Intelligence*. Springer. 10.1007/978-3-030-26354-6_32

Industrial Skyworks. (n.d.). *Drone inspection services*. https://industrialskyworks.com/droneinspections-services

Jabbar, R., Dhib, E., Said, A. B., Krichen, M., Fetais, N., Zaidan, E., & Barkaoui, K. (2022). Blockchain Technology for Intelligent Transportation Systems: A Systematic Literature Review. *IEEE Access: Practical Innovations, Open Solutions, 10*, 20995–21031. doi:10.1109/ACCESS.2022.3149958

Janakbhai, N. D., Saurin, M. J., & Patel, M. (2021). Blockchain-Based Intelligent Transportation System with Priority Scheduling. In Data Science and Intelligent Applications. Lecture Notes on Data Engineering and Communications Technologies (vol. 52, pp. 311-317). Springer. doi:10.1007/978-981-15-4474-3_34

Jang, J., Ahn, T., & Han, J. (2017). A new application-layer overlay platform for better connected vehicles. *International Journal of Distributed Sensor Networks, 13*(11). Advance online publication. doi:10.1177/1550147717742072

Jasim, M. M., Al-Qaysi, H. K., Allbadi, Y., & Al-Azzawi, H. M. (2020). Comprehensive study on unmanned aerial vehicles (UAVs). Advanced Mathematical Models & Applications, 5(2), 240-259.

Javaid, M., Haleem, A., Singh, R. P., Khan, S., & Suman, R. (2021). Blockchain technology applications for Industry 4.0: A literature-based review. *Blockchain: Research and Applications, 2*(4), 100027. doi:10.1016/j.bcra.2021.100027

Jayapal, C., & Roy, S. S. (2016). Road traffic congestion management using VANET. In *International conference on advances in human machine interaction (HMI)* (pp. 1-7). IEEE.

Jensen, C. S., & Dyreson, C. E. (1998). The consensus glossary of temporal database concepts – February 1998 version. In Temporal Databases – Research and Practice (pp. 367–405). Berlin, Germany: Springer.

Jensen, C. S., & Snodgrass, R. T. (2018a). Temporal Database. In L. Liu & M. T. Özsu (Eds.), *Encyclopedia of Database Systems* (2nd ed.). Springer-Verlag. doi:10.1007/978-1-4614-8265-9_1415

Jensen, C. S., & Snodgrass, R. T. (2018b). Transaction Time. In L. Liu & M. T. Özsu (Eds.), *Encyclopedia of Database Systems*. Springer. doi:10.1007/978-1-4614-8265-9_1064

Jensen, C. S., & Snodgrass, R. T. (2018c). Valid Time. In L. Liu & M. T. Özsu (Eds.), *Encyclopedia of Database Systems*. Springer. doi:10.1007/978-1-4614-8265-9_1066

Jensen, C. S., & Snodgrass, R. T. (2018d). Bitemporal Relation. In L. Liu & M. T. Özsu (Eds.), *Encyclopedia of Database Systems*. Springer. doi:10.1007/978-1-4614-8265-9_1409

Jensen, C. S., & Snodgrass, R. T. (2018f). Temporal Query Languages. In L. Liu & M. T. Özsu (Eds.), *Encyclopedia of Database Systems*. Springer. doi:10.1007/978-1-4614-8265-9_407

Jian, L., Li, Z., Yang, X., Wu, W., Ahmad, A., & Jeon, G. (2019). Combining unmanned aerial vehicles with artificial-intelligence technology for traffic-congestion recognition: electronic eyes in the skies to spot clogged roads. *IEEE Consumer Electronics Magazine*, 81-86.

Jiang, J., & Han, G. (2018). Routing protocols for unmanned aerial vehicles. *IEEE Communications Magazine*, *56*(1), 58–63. doi:10.1109/MCOM.2017.1700326

Jiao, Z., Zhang, Y., Xin, J., Mu, L., Yi, Y., & Liu, H. (2019). A deep learning based forest fire detection approach using UAV and YOLOv3. In *1st International conference on industrial artificial intelligence (IAI)* (pp. 1--5). IEEE. 10.1109/ICIAI.2019.8850815

Kakamoukas, G. A., Sarigiannidis, P. G., & Economides, A. A. (2020). FANETs in Agriculture-A routing protocol survey. *Internet of Things*, 100183.

Kaleem, Z., & Rehmani, M. H. (2018). Amateur drone monitoring: State-of-the-art architectures, key enabling technologies, and future research directions. *IEEE Wireless Communications*, *25*(2), 150–159. doi:10.1109/MWC.2018.1700152

Kang, J., Yu, R., Huang, X., Jonsson, M., Bogucka, H., Gjessing, S., & Zhang, Y. (2016). Location privacy attacks and defenses in cloud-enabled internet of vehicles. *IEEE Wireless Communications*, *23*(5), 52–59. doi:10.1109/MWC.2016.7721742

Kazi, A. K., Khan, S. M., & Haider, N. G. (2021). Reliable group of vehicles (RGoV) in VANET. *IEEE Access: Practical Innovations, Open Solutions*, *9*, 111407–111416. doi:10.1109/ACCESS.2021.3102216

Kazmi, W., Bisgaard, M., Garcia-Ruiz, F., Hansen, K. D., & la Cour-Harbo, A. (2011). Adaptive surveying and early treatment of crops with a team of autonomous vehicles. In *Proceedings of the 5th European Conference on Mobile Robots ECMR 2011* (pp. 253-258). Academic Press.

Ke, R., Li, Z., Kim, S., Ash, J., Cui, Z., & Wang, Y. (2016). Real-time bidirectional traffic flow parameter estimation from aerial videos. *IEEE Transactions on Intelligent Transportation Systems*, *18*(4), 890–901. doi:10.1109/TITS.2016.2595526

Khan, M. A., Safi, A., Qureshi, I. M., & Khan, I. U. (2017, November). Flying ad-hoc networks (FANETs): A review of communication architectures, and routing protocols. In *2017 First international conference on latest trends in electrical engineering and computing technologies (INTELLECT)* (pp. 1-9). IEEE.

Khan, I. U., Qureshi, I. M., Aziz, M. A., Cheema, T. A., & Shah, S. B. H. (2020). Smart IoT control-based nature inspired energy efficient routing protocol for flying ad hoc network (FANET). *IEEE Access: Practical Innovations, Open Solutions*, *8*, 56371–56378. doi:10.1109/ACCESS.2020.2981531

Khan, K. M., Arshad, J., & Khan, M. M. (2020). Investigating performance constraints for blockchain based secure e-voting system. *Future Generation Computer Systems*, *105*, 13–26. doi:10.1016/j.future.2019.11.005

Khan, M. A., Khan, I. U., Safi, A., & Quershi, I. M. (2018). Dynamic routing in flying ad-hoc networks using topology-based routing protocols. *Drones*, *2*(3), 27. doi:10.3390/drones2030027

Khan, M. A., Qureshi, I. M., & Khanzada, F. (2019). A hybrid communication scheme for efficient and low-cost deployment of future flying ad-hoc network (FANET). *Drones (Basel)*, *3*(1), 16. doi:10.3390/drones3010016

Khan, N. A., Jhanjhi, N., Brohi, S. N., Usmani, R. S., & Nayyar, A. (2020). Smart traffic monitoring system using unmanned aerial vehicles (UAVs). *Computer Communications*, *157*, 434–443. doi:10.1016/j.comcom.2020.04.049

Kheli, F., Bradai, A., Singh, K., & Atri, M. (2018). Localization and Energy-Efficient Data Routing for Unmanned Aerial Vehicles: Fuzzy-Logic-Based Approach. *IEEE Communications Magazine*, *56*(4), 129–133. doi:10.1109/MCOM.2018.1700453

Kher, R., Terjesen, S., & Liu, C. (2020). Blockchain, Bitcoin, and ICOs: A review and research agenda. *Small Business Economics*, *56*(4), 1699–1720. doi:10.100711187-019-00286-y

Khuwaja, A. A., Chen, Y., Zhao, N., Alouini, M. S., & Dobbins, P. (2018). A survey of channel modeling for UAV communications. *IEEE Communications Surveys and Tutorials*, *20*(4), 2804–2821. doi:10.1109/COMST.2018.2856587

Kim, S. (2018). Blockchain for a trust network among intelligent vehicles. []. Elsevier.]. *Advances in Computers*, *111*, 43–68. doi:10.1016/bs.adcom.2018.03.010

Krichen, L., Fourati, M., & Fourati, L. C. (2018, September). Communication architecture for unmanned aerial vehicle system. In *International Conference on Ad-Hoc Networks and Wireless* (pp. 213-225). Springer. 10.1007/978-3-030-00247-3_20

Krishnan & Kumar. (2020). Security and Privacy in VANET: Concepts, Solutions and Challenges. *2020 International Conference on Inventive Computation Technologies (ICICT)*, 789-794. doi: 10.1109/ICICT48043.2020.9112535

Kulkarni, K., & Michels, J. E. (2012). Temporal features in SQL: 2011. *SIGMOD Record*, *41*(3), 34–43. doi:10.1145/2380776.2380786

Kulkarni, S., Chaphekar, V., Chowdhury, M. M., Erden, F., & Guvenc, I. (2020). UAV aided search and rescue operation using reinforcement learning. In *SoutheastCon* (pp. 1–8). IEEE. doi:10.1109/SoutheastCon44009.2020.9368285

Kumar, V., Mishra, S., & Chand, N. (2013). Applications of VANETs: present & future. *Communications and Network*, 12.

Kumari, K., Sah, B., & Maakar, S. (2015). A survey: Different mobility model for FANET. *International Journal of Advanced Research in Computer Science and Software Engineering*.

Kumar, T. P., & Krishna, P. V. (2018). Power modelling of sensors for IoT using reinforcement learning. *International Journal of Advanced Intelligence Paradigms*, *10*(1-2), 3–22.

Kumar, T. P., & Krishna, P. V. (2021). A survey of energy modeling and efficiency techniques of sensors for IoT systems. *International Journal of Sensors, Wireless Communications and Control*, *11*(3), 271–283. doi:10.2174/2210327910999200614001521

Kumbhar, F. H., & Shin, S. Y. (2020). DT-VAR: Decision tree predicted compatibility-based vehicular ad-hoc reliable routing. *IEEE Wireless Communications Letters*, *10*(1), 87–91. doi:10.1109/LWC.2020.3021430

Lakew, D. S., Sa'ad, U., Dao, N. N., Na, W., & Cho, S. (2020). Routing in flying ad hoc networks: A comprehensive survey. *IEEE Communications Surveys and Tutorials*, *22*(2), 1071–1120. doi:10.1109/COMST.2020.2982452

Lee, E.-K., Gerla, M., Pau, G., Lee, U., & Lim, J.-H. (2016). Internet of vehicles: From intelligent grid to autonomous cars and vehicular fogs. *International Journal of Distributed Sensor Networks*, *12*(9). doi:10.1177/1550147716665500

Lee, U., Lee, J., Park, J.-S., & Gerla, M. (2009). FleaNet: A virtual market place on vehicular networks. *IEEE Transactions on Vehicular Technology*, 344–355.

Li, J., Zhou, Y., & Lamont, L. (2013). Communication architectures and protocols for networking unmanned aerial vehicles. In IEEE Globecom Workshops (GC Wkshps) (pp. 1415-1420). IEEE.

Liang, W., Li, Z., Zhang, H., Wang, S., & Bie, R. (2015). Vehicular ad hoc networks: Architectures, research issues, methodologies, challenges, and trends. *International Journal of Distributed Sensor Networks*, *11*(8), 745303. doi:10.1155/2015/745303

Li, C., Fu, Y., Yu, F. R., Luan, T. H., & Zhang, Y. (2020a). Vehicle position correction: A vehicular blockchain networks-based GPS error sharing framework. *IEEE Transactions on Intelligent Transportation Systems*, *22*(2), 898–912. doi:10.1109/TITS.2019.2961400

Lin, N., Gao, F., Zhao, L., Al-Dubai, A., & Tan, Z. (2019, August). A 3D smooth random walk mobility model for FANETs. In *2019 IEEE 21st International Conference on High Performance Computing and Communications; IEEE 17th International Conference on Smart City; IEEE 5th International Conference on Data Science and Systems (HPCC/SmartCity/DSS)* (pp. 460-467). IEEE.

Lin, Y. J., & Sun, M. T. (2018, October). HyperQL-Efficient Blockchain Query. In *Proc. of the 42th Taiwan Academic Network Conference (TANET)* (pp. 939-944). Academic Press.

Lin, J., Cai, W., Zhang, S., Fan, X., Guo, S., & Dai, J. (2018). A Survey of Flying Ad-Hoc Networks: Characteristics and Challenges. In *Eighth International Conference on Instrumentation & Measurement, Computer, Communication and Control (IMCCC)* (pp. 766--771). IEEE. 10.1109/IMCCC.2018.00165

Liu, B., Jia, D., Wang, J., Lu, K., & Wu, L. (2015). Cloud-assisted safety message dissemination in VANET–cellular heterogeneous wireless network. *IEEE Systems Journal, 11*(1), 128-139.

Liu, Y., Niu, J., Qu, G., Cai, Q., & Ma, J. (2011). Message delivery delay analysis in VANETs with a bidirectional traffic model. *2011 7th International Wireless Communications and Mobile Computing Conference,* 1754-1759. doi: 10.1109/IWCMC.2011.5982801

Liu, J., Peng, S., Long, C., Wei, L., Liu, Y., & Tian, Z. (2020, March). Blockchain for Data Science. In *Proceedings of the 2020 The 2nd International Conference on Blockchain Technology* (pp. 24-28). 10.1145/3390566.3391681

Li, W., Wu, J., Cao, J., Chen, N., Zhang, Q., & Buyya, R. (2021). Blockchain-based trust management in cloud computing systems: A taxonomy, review and future directions. *Journal of Cloud Computing: Advances. Systems and Applications, 10*(1), 1–34.

Li, X., Jiang, P., Chen, T., Luo, X., & Wen, Q. (2020b). A survey on the security of blockchain systems. *Future Generation Computer Systems, 107*, 841–853. doi:10.1016/j.future.2017.08.020

Li, Y., & Cai, L. (2017). UAV-assisted dynamic coverage in a heterogeneous cellular system. *IEEE Network, 31*(4), 56–61. doi:10.1109/MNET.2017.1600280

Li, Y., Zheng, K., Yan, Y., Liu, Q., & Zhou, X. (2017, March). EtherQL: a query layer for blockchain system. In *International Conference on Database Systems for Advanced Applications (DASFAA)* (pp. 556-567). Springer. 10.1007/978-3-319-55699-4_34

Louargant, M., Villette, S., Jones, G., Vigneau, N., Paoli, J. N., & Gée, C. (2017). Weed detection by UAV: Simulation of the impact of spectral mixing in multispectral images. *Precision Agriculture, 18*(6), 932–951. doi:10.100711119-017-9528-3

Lu, J., Wan, S., Chen, X., Chen, Z., Fan, P., & Letaief, K. B. (2018). Beyond empirical models: Pattern formation driven placement of UAV base stations. *IEEE Transactions on Wireless Communications, 17*(6), 3641–3655. doi:10.1109/TWC.2018.2812167

Luo, G., Li, J., Zhang, L., Yuan, Q., Liu, Z., & Yang, F. (2018). sdnMAC: A software-defined network inspired MAC protocol for cooperative safety in VANETs. *IEEE Transactions on Intelligent Transportation Systems*, *19*(6), 2011–2024. doi:10.1109/TITS.2017.2736887

Maaroufi, S., & Pierre, S. (2021). BCOOL: A novel blockchain congestion control architecture using dynamic service function chaining and machine learning for next generation vehicular networks. *IEEE Access: Practical Innovations, Open Solutions*, *9*, 53096–53122. doi:10.1109/ACCESS.2021.3070023

Macrina, G., Pugliese, L. D., Guerriero, F., & Laporte, G. (2020). Drone-aided routing: A literature review. *Transportation Research Part C, Emerging Technologies*, *120*, 102762. doi:10.1016/j.trc.2020.102762

Maddio, S., Cidronali, A., Palonghi, A., & Manes, G. (2013, June). A reconfigurable leakage canceler at 5.8 GHz for DSRC applications. In *2013 IEEE MTT-S International Microwave Symposium Digest (MTT)* (pp. 1-3). IEEE.

Ma, G., Ge, C., & Zhou, L. (2020). Achieving reliable timestamp in the bitcoin platform. *Peer-to-Peer Networking and Applications*, *13*(6), 2251–2259. doi:10.100712083-020-00905-6

Marconato, E. A., Maxa, J. A., Pigatto, D. F., Pinto, A. S., Larrieu, N., & Branco, K. R. C. (2016, June). IEEE 802.11 n vs. IEEE 802.15. 4: A study on communication QoS to provide safe FANETs. In *2016 46th Annual IEEE/IFIP international conference on dependable systems and networks workshop (DSN-W)* (pp. 184-191). IEEE.

Mathur, P., Nielsen, R. H., Prasad, N. R., & Prasad, R. (2016). Data collection using miniature aerial vehicles in wireless sensor networks. *IET Wireless Sensor Systems*, *6*(1), 17–25. doi:10.1049/iet-wss.2014.0120

Maxa, J. A., Mahmoud, M. S. B., & Larrieu, N. (2017). Survey on UAANET routing protocols and network security challenges. *Ad-Hoc & Sensor Wireless Networks*, 37.

McTrans. (2022, August 18). *McTrans center*. https://mctrans.ce.ufl.edu/

Meneguette, R. I., Bittencourt, L. F., & Madeira, E. R. M. (2013). A seamless flow mobility management architecture for vehicular communication networks. *Journal of Communications and Networks (Seoul)*, *15*(2), 207–216. doi:10.1109/JCN.2013.000034

Modi, Y., Panchal, M., Bhatia, J., & Tanwar, S. (2022). Blockchain-Based Software-Defined Vehicular Networks for Intelligent Transportation System Beyond 5G. In *Advances in Computing, Informatics, Networking and Cybersecurity* (pp. 513–533). Springer. doi:10.1007/978-3-030-87049-2_17

Mohan, C. (2018, April). Blockchains and databases: A new era in distributed computing. In *2018 IEEE 34th International Conference on Data Engineering (ICDE)* (pp. 1739-1740). IEEE. 10.1109/ICDE.2018.00227

Mohanty, A., Mahapatra, S., & Bhanja, U. (2019). Traffic congestion detection in a city using clustering techniques in VANETs. *Indonesian Journal of Electrical Engineering and Computer Science*, 884-891.

Monir, N., Toraya, M. M., Vladyko, A., Muthanna, A., Torad, M. A., El-Samie, F. E. A., & Ateya, A. A. (2022). Seamless Handover Scheme for MEC/SDN-Based Vehicular Networks. *Journal of Sensor and Actuator Networks*, *11*(1), 9. doi:10.3390/jsan11010009

Motlagh, N. H., Bagaa, M., & Taleb, T. (2017). UAV-based IoT platform: A crowd surveillance use case. *IEEE Communications Magazine*, *55*(2), 128–134. doi:10.1109/MCOM.2017.1600587CM

Motlagh, N. H., Taleb, T., & Arouk, O. (2016). Low-altitude unmanned aerial vehicles-based internet of things services: Comprehensive survey and future perspectives. *IEEE Internet of Things Journal*, *3*(6), 899–922. doi:10.1109/JIOT.2016.2612119

Moustafa, H., & Zhang, Y. (2009). *Vehicular networks: Techniques, Standards, and Applications*. CRC Press.

Mozaffari, M., Saad, W., Bennis, M., Nam, Y. H., & Debbah, M. (2019). A tutorial on UAVs for wireless networks: Applications, challenges, and open problems. *IEEE Communications Surveys and Tutorials*, *21*(3), 2334–2360. doi:10.1109/COMST.2019.2902862

Muchiri, N., & Kimathi, S. (2016, June). A review of applications and potential applications of UAV. In *Proceedings of sustainable research and innovation conference* (pp. 280-283). Academic Press.

Mukhopadhyay, A. (2020). FANET based Emergency Healthcare Data Dissemination. In *Second International Conference on Inventive Research in Computing Applications (ICIRCA)* (pp. 170-175). IEEE.

Mukhopadhyay, A., & Raghunath, S. (2016). Feasibility and performance evaluation of VANET techniques to enhance real-time emergency healthcare services. In *International Conference on Advances in Computing, Communications and Informatics (ICACCI)* (pp. 2597--2603). IEEE. 10.1109/ICACCI.2016.7732449

Muniyandi, R.,, Qamar, F.,, & Jasim, Naeem, A. (2020). Genetic Optimized Location Aided Routing Protocol for VANET Based on Rectangular Estimation of Position. *Applied Sciences (Basel, Switzerland)*, *10*, 5759. doi:10.3390/app10175759

Nakamoto, S. (2008). *Bitcoin: A peer-to-peer electronic cash system*. https://bitcoin.org/bitcoin.pdf

Nampally, V., & Sharma, M. R. (2018). Information sharing standards in communication for VANET. *International Journal of Scientific Research in Computer Science Applications and Management Studies*, 2319–1953.

Narbayeva, S., Bakibayev, T., Abeshev, K., Makarova, I., Shubenkova, K., & Pashkevich, A. (2020). Blockchain Technology on the Way of Autonomous Vehicles Development. *Transportation Research Procedia*, *44*, 168–175. doi:10.1016/j.trpro.2020.02.024

Nasrulin, B., Muzammal, M., & Qu, Q. (2018, November). A robust spatio-temporal verification protocol for blockchain. In *International Conference on Web Information Systems Engineering (WISE)* (pp. 52-67). Springer. 10.1007/978-3-030-02922-7_4

Navulur, S., & Prasad, M. G. (2017). Agricultural management through wireless sensors and internet of things. *Iranian Journal of Electrical and Computer Engineering*, *7*(6), 3492. doi:10.11591/ijece.v7i6.pp3492-3499

Ning, Z., Hu, X., Chen, Z., Zhou, M., Hu, B., Cheng, J., & Obaidat, M. S. (2017). A cooperative quality-aware service access system for social internet of vehicles. *IEEE Internet of Things Journal*, *5*(4), 2506–2517. doi:10.1109/JIOT.2017.2764259

Ning, Z., Sun, S., Wang, X., Guo, L., Guo, S., Hu, X., Hu, B., & Kwok, R. (2021). Blockchain-enabled intelligent transportation systems: A distributed crowdsensing framework. *IEEE Transactions on Mobile Computing*, 1. Advance online publication. doi:10.1109/TMC.2021.3079984

Nofer, M., Gomber, P., Hinz, O., & Schiereck, D. (2017). Blockchain. *Business & Information Systems Engineering*, *59*(3), 183–187. doi:10.100712599-017-0467-3

Oche, M., Tambuwal, A. B., Chemebe, C., Noor, R. M., & Distefano, S. (2020). VANETs QoS-based routing protocols based on multi-constrained ability to support ITS infotainment services. *Wireless Networks*, *26*(3), 1685–1715. doi:10.100711276-018-1860-7

Orsino, A., Ometov, A., Fodor, G., Moltchanov, D., Militano, L., Andreev, S., Yilmaz, O. N. C., Tirronen, T., Torsner, J., Araniti, G., Iera, A., Dohler, M., & Koucheryavy, Y. (2017). Effects of heterogeneous mobility on D2D-and drone-assisted mission-critical MTC in 5G. *IEEE Communications Magazine*, *55*(2), 79–87. doi:10.1109/MCOM.2017.1600443CM

Oubbati, O. S., Lakas, A., Zhou, F., G✓neş, M., & Yagoubi, M. B. (2017). A survey on position-based routing protocols for Flying Ad hoc Networks (FANETs). *Vehicular Communications*, *10*, 29–56. doi:10.1016/j.vehcom.2017.10.003

Park, S. Y., Shin, C. S., Jeong, D., & Lee, H. (2018). DroneNetX: Network reconstruction through connectivity probing and relay deployment by multiple UAVs in ad hoc networks. *IEEE Transactions on Vehicular Technology*, *67*(11), 11192–11207. doi:10.1109/TVT.2018.2870397

Paul, A., Daniel, A., Ahmad, A., & Rho, S. (2015). Cooperative cognitive intelligence for internet of vehicles. *IEEE Systems Journal*, *11*(3), 1249–1258. doi:10.1109/JSYST.2015.2411856

Pavithran, D., Shaalan, K., Al-Karaki, J. N., & Gawanmeh, A. (2020). Towards building a blockchain framework for IoT. *Cluster Computing*, *23*(3), 2089–2103. doi:10.100710586-020-03059-5

Pavithra, T., & Nagabhushana, B. S. (2020). A Survey on Security in VANETs. *2020 Second International Conference on Inventive Research in Computing Applications (ICIRCA)*, 881-889. doi: 10.1109/ICIRCA48905.2020.9182823

Poonia, R. C. (2018). A performance evaluation of routing protocols for vehicular ad hoc networks with swarm intelligence. *International Journal of System Assurance Engineering and Management*, *9*(4), 830–835. doi:10.100713198-017-0661-1

Pratama, F. A., & Mutijarsa, K. (2018, October). Query Support for Data Processing and Analysis on Ethereum Blockchain. In *2018 International Symposium on Electronics and Smart Devices (ISESD)* (pp. 1-5). IEEE. 10.1109/ISESD.2018.8605476

Primicerio, J., Di Gennaro, S. F., Fiorillo, E., Genesio, L., Lugato, E., Matese, A., & Vaccari, F. P. (2012). A flexible unmanned aerial vehicle for precision agriculture. *Precision Agriculture, 13*(4), 517–523. doi:10.100711119-012-9257-6

PwC. (n.d.). *Emerging technology*. http://usblogs.pwc.com/emergingtechnology/

Qu, Q., Nurgaliev, I., Muzammal, M., Jensen, C. S., & Fan, J. (2019). On spatio-temporal blockchain query processing. *Future Generation Computer Systems, 98*, 208–218. doi:10.1016/j.future.2019.03.038

Rajan, J., Shriwastav, S., Kashyap, A., Ratnoo, A., & Ghose, D. (2021). Disaster management using unmanned aerial vehicles. In *Unmanned Aerial Systems* (pp. 129–155). Academic Press. doi:10.1016/B978-0-12-820276-0.00013-3

Redstagfulfillment. (n.d.a). *The future of distribution*. https://redstagful_llment.com/the-future-ofdistribution/

Redstagfulfillment. (n.d.b). https://redstagful_llment.com

Rehman, M., Khan, Z. A., Javed, M. U., Iftikhar, M. Z., Majeed, U., Bux, I., & Javaid, N. (2020, April). A Blockchain Based Distributed Vehicular Network Architecture for Smart Cities. In *Workshops of the International Conference on Advanced Information Networking and Applications* (pp. 320-331). Springer. 10.1007/978-3-030-44038-1_29

Reshma, R., Ramesh, T., & Sathishkumar, P. (2016, January). Security situational aware intelligent road traffic monitoring using UAVs. In *2016 international conference on VLSI systems, architectures, technology and applications (VLSI-SATA)* (pp. 1-6). IEEE.

Saleem, M. A., Shijie, Z., & Sharif, A. (2019). Data transmission using iot in vehicular ad-hoc networks in smart city congestion. *Mobile Networks and Applications, 24*(1), 248–258. doi:10.100711036-018-1205-x

Salvo, P., Felice, M. D., Cuomo, F., & Baiocchi, A. (2012). Infotainment traffic flow dissemination in an urban VANET. In *IEEE Global Communications Conference (GLOBECOM)* (pp. 67-72). IEEE. 10.1109/GLOCOM.2012.6503092

Sami Oubbati, O. C., Chaib, N., Lakas, A., Bitam, S., & Lorenz, P. (2020). U2RV: UAV-assisted reactive routing protocol for VANETs. *International Journal of Communication Systems, 33*(10), 4104. doi:10.1002/dac.4104

Sánchez-García, J., García-Campos, J. M., Arzamendia, M., Reina, D. G., Toral, S. L., & Gregor, D. (2018). A survey on unmanned aerial and aquatic vehicle multi-hop networks: Wireless communications, evaluation tools and applications. *Computer Communications, 119*, 43–65. doi:10.1016/j.comcom.2018.02.002

Sankarasrinivasan, S., Balasubramanian, E., Karthik, K., Chandrasekar, U., & Gupta, R. (2015). Health monitoring of civil structures with integrated UAV and image processing system. *Procedia Computer Science*, *54*, 508–515. doi:10.1016/j.procs.2015.06.058

Sarakis, L., Orphanoudakis, T., Leligou, H. C., Voliotis, S., & Voulkidis, A. (2016). Providing entertainment app lications in VANET environments. *IEEE Wireless Communications*, *23*(1), 30–37. doi:10.1109/MWC.2016.7422403

Sawalmeh, A. H., & Othman, N. S. (2021). *An overview of collision avoidance approaches and network architecture of unmanned aerial vehicles (UAVs).* arXiv preprint arXiv:2103.14497.

Seliem, H., Shahidi, R., Ahmed, M. H., & Shehata, M. S. (2018). Drone-based highway-VANET and DAS service. *IEEE Access: Practical Innovations, Open Solutions*, *6*, 20125–20137. doi:10.1109/ACCESS.2018.2824839

Senapati, B. R. (2021). Composite fault diagnosis methodology for urban vehicular ad hoc network. *Vehicular Communications*, 100337.

Senapati, B. R., & Khilar, P. M. (2020). Automatic parking service through VANET: A convenience application. In Progress in Computing, Analytics and Networking (pp. 151-159). Springer. doi:10.1007/978-981-15-2414-1_16

Senapati, B. R., & Khilar, P. M. (2020). Optimization of performance parameter for vehicular ad-hoc network (VANET) using swarm intelligence. In Nature Inspired Computing for Data Science (pp. 83-107). Springer.

Senapati, B. R., Khilar, P. M., & Sabat, N. K. (2019). An automated toll gate system using vanet. In *IEEE 1st international conference on energy, systems and information processing (ICESIP)* (pp. 1-5). IEEE.

Senapati, B. R., Khilar, P. M., & Swain, R. R. (2021). Composite fault diagnosis methodology for urban vehicular ad hoc network. *Vehicular Communications*, 100337.

Senapati, B. R., Khilar, P. M., & Swain, R. R. (2021). Environmental monitoring through Vehicular Ad Hoc Network: A productive application for smart cities. *International Journal of Communication Systems*, 4988.

Senapati, B. R., Khilar, P. M., & Swain, R. R. (2021). Fire controlling under uncertainty in urban region using smart vehicular ad hoc network. *Wireless Personal Communications*, *116*(3), 2049–2069. doi:10.100711277-020-07779-0

Senapati, B. R., Mohapatra, S., & Khilar, P. M. (2021). Fault Detection for VANET Using Vehicular Cloud. In *Intelligent and Cloud Computing* (pp. 87–95). Springer. doi:10.1007/978-981-15-6202-0_10

Senapati, B. R., Swain, R. R., & Khilar, P. M. (2020). Environmental monitoring under uncertainty using smart vehicular ad hoc network. In *Smart intelligent computing and applications* (pp. 229–238). Springer. doi:10.1007/978-981-13-9282-5_21

Senouci, O., Aliouat, Z., & Harous, S. (2019). A review of routing protocols in internet of vehicles and their challenges. *Sensor Review*, *39*(1), 58–70. doi:10.1108/SR-08-2017-0168

Shah, A. S., Karabulut, M. A., Ilhan, H., & Tureli, U. (2020). Performance optimization of cluster-based MAC protocol for VANETs. *IEEE Access: Practical Innovations, Open Solutions*, *8*, 167731–167738. doi:10.1109/ACCESS.2020.3023642

Shahzad, M., & Antoniou, J. (2019). Quality of user experience in 5G-VANET. In *IEEE 24th international workshop on computer aided modeling and design of communication links and networks (camad)* (pp. 1-6). IEEE.

Sharma, V., & Kumar, R. (2017). Cooperative frameworks and network models for flying ad hoc networks: A survey. *Concurrency and Computation*, *29*(4), e3931. doi:10.1002/cpe.3931

Sharma, V., Kumar, R., & Kumar, N. (2018). DPTR: Distributed priority tree-based routing protocol for FANETs. *Computer Communications*, *122*, 129–151. doi:10.1016/j.comcom.2018.03.002

Sharples, M., & Domingue, J. (2016, September). The blockchain and kudos: A distributed system for educational record, reputation and reward. In *European conference on technology enhanced learning* (pp. 490-496). Springer. 10.1007/978-3-319-45153-4_48

Sheikh, M. S., & Liang, J. (2019). A comprehensive survey on VANET security services in traffic management system. *Wireless Communications and Mobile Computing*, *2019*, 1–23. doi:10.1155/2019/2423915

Singh, A., Parizi, R. M., Zhang, Q., Choo, K. K. R., & Dehghantanha, A. (2020). Blockchain smart contracts formalization: Approaches and challenges to address vulnerabilities. *Computers & Security*, *88*, 101654. doi:10.1016/j.cose.2019.101654

Singh, K., & Verma, A. K. (2019). Flying adhoc networks concept and challenges. In *Advanced methodologies and technologies in network architecture, mobile computing, and data analytics* (pp. 903–911). IGI Global.

Singh, S. K. (2015). A comprehensive survey on fanet: Challenges and advancements. *International Journal of Computer Science and Information Technologies*, 2010–2013.

Singh, S. K. (2015). A comprehensive survey on FANET: Challenges and advancements. *International Journal of Computer Science and Information Technologies*, *6*(3), 2010–2013.

Six, N., Herbaut, N., & Salinesi, C. (2022). Blockchain software patterns for the design of decentralized applications: A systematic literature review. *Blockchain: Research and Applications*, *3*(2), 100061. doi:10.1016/j.bcra.2022.100061

Skorobogatov, G., Barrado, C., & Salamí, E. (2020). Multiple UAV systems: A survey. *Unmanned Systems*, *8*(02), 149–169. doi:10.1142/S2301385020500090

Snodgrass. (Ed.). (1995). *The TSQL2 Temporal Query Language*. Kluwer Academic Publishers.

Srinivasakumar, V., Vanamoorthy, M., Sairaj, S., & Ganesh, S. (2022). An alternative C++-based HPC system for Hadoop MapReduce. *Open Computer Science*, *12*(1), 238–247. doi:10.1515/comp-2022-0246

Srivastava, A., & Prakash, J. (2021). Future FANET with application and enabling techniques: Anatomization and sustainability issues. *Computer Science Review*, *39*, 100359. doi:10.1016/j.cosrev.2020.100359

Sudhakar, S., Vijayakumar, V., Kumar, C. S., Priya, V., Ravi, L., & Subramaniyaswamy, V. (2020). Unmanned Aerial Vehicle (UAV) based Forest Fire Detection and monitoring for reducing false alarms in forest-fires. *Computer Communications*, *149*, 1–16. doi:10.1016/j.comcom.2019.10.007

Su, H., & Zhang, X. (2007). Clustering-based multichannel MAC protocols for QoS provisionings over vehicular ad hoc networks. *IEEE Transactions on Vehicular Technology*, *56*(6), 3309–3323.

Sumi, L., & Ranga, V. (2018). *An iot-vanet-based traffic management system for emergency vehicles in a smart city. In Recent Findings in Intelligent Computing Techniques*. Springer.

Sun, W. L., Liu, J., & Zhang, H. (2017). When smart wearables meet intelligent vehicles: Challenges and future directions. *IEEE Wireless Communications*, *24*(3), 58–65. doi:10.1109/MWC.2017.1600423

Sutheerakul, C., Kronprasert, N., Kaewmoracharoen, M., & Pichayapan, P. (2017). Application of unmanned aerial vehicles to pedestrian traffic monitoring and management for shopping streets. *Transportation Research Procedia*, *25*, 1717–1734. doi:10.1016/j.trpro.2017.05.131

Swain, K. C., Thomson, S. J., & Jayasuriya, H. P. (2010). Adoption of an unmanned helicopter for low-altitude remote sensing to estimate yield and total biomass of a rice crop. *Transactions of the ASABE*, *53*(1), 21–27. doi:10.13031/2013.29493

Swain, S., Khilar, P. M., & Senapati, B. R. (2022). An effective data routing for dynamic area coverage using multidrone network. *Transactions on Emerging Telecommunications Technologies*, *33*(9), 4532. doi:10.1002/ett.4532

Szalachowski, P. (2018, June). (Short Paper) Towards More Reliable Bitcoin Timestamps. In *2018 Crypto Valley Conference on Blockchain Technology (CVCBT)* (pp. 101-104). IEEE. 10.1109/CVCBT.2018.00018

Taeihagh, A., & Lim, H. S. M. (2019). Governing autonomous vehicles: Emerging responses for safety, liability, privacy, cybersecurity, and industry risks. *Transport Reviews*, *39*(1), 103–128. doi:10.1080/01441647.2018.1494640

Tansel, A. U. (1997). Temporal relational data model. *IEEE Transactions on Knowledge and Data Engineering*, *9*(3), 464–479. doi:10.1109/69.599934

Tareque, M. H., Hossain, M. S., & Atiquzzaman, M. (2015, September). On the routing in flying ad hoc networks. In *2015 federated conference on computer science and information systems (FedCSIS)* (pp. 1-9). IEEE.

Tariq, R. I., Iqbal, Z., & Aadil, F. (2020). IMOC: Optimization technique for drone-assisted VANET (DAV) based on moth flame optimization. *Wireless Communications and Mobile Computing, 2020*, 860646. doi:10.1155/2020/8860646

Temel & Bekmezci. (2015). LODMAC: Location Oriented Directional MAC Protocol for FANETs. *Computer Networks, 83*, 76–84. . doi:10.1016/j.comnet.2015.03.001

Tetcos. (n.d.). *Tetcos: NetSim - Network Simulation Software, India*. https://www.tetcos.com/

Thammawichai, M., Baliyarasimhuni, S. P., Kerrigan, E. C., & Sousa, J. B. (2017). Optimizing communication and computation for multi-UAV information gathering applications. *IEEE Transactions on Aerospace and Electronic Systems*, *54*(2), 601–615. doi:10.1109/TAES.2017.2761139

Tiennoy, S., & Saivichit, C. (2018). Using a distributed roadside unit for the data dissemination protocol in VANET with the named data architecture. *IEEE Access: Practical Innovations, Open Solutions*, *6*, 32612–32623. doi:10.1109/ACCESS.2018.2840088

Tolba, A. (2019). Content accessibility preference approach for improving service optimality in internet of vehicles. *Computer Networks*, *152*, 78–86. doi:10.1016/j.comnet.2019.01.038

Tomescu, A., & Devadas, S. (2017, May). Catena: Efficient non-equivocation via bitcoin. In *2017 IEEE Symposium on Security and Privacy (SP)* (pp. 393-409). IEEE. 10.1109/SP.2017.19

Toor, Miihlethaler, Laouiti, & De La Fortelle. (2008). Vehicle Ad Hoc Networks: Applications and Related Technical issues. *IEEE Communications Surveys & Tutorials, 10*(3), 74-88.

Toutouh, J., & Alba, E. (2018). A swarm algorithm for collaborative traffic in vehicular networks. *Vehicular Communications*, *12*, 127–137. doi:10.1016/j.vehcom.2018.04.003

Tran, T. X., Hajisami, A., & Pompili, D. (2017). Cooperative Hierarchical Caching in 5G Cloud Radio Access Networks. *IEEE Network*, *31*(4), 35–41. doi:10.1109/MNET.2017.1600307

Trotta, A., Di Felice, M., Montori, F., Chowdhury, K. R., & Bononi, L. (2018). Joint coverage, connectivity, and charging strategies for distributed UAV networks. *IEEE Transactions on Robotics*, *34*(4), 883–900. doi:10.1109/TRO.2018.2839087

Tschorsch, F., & Scheuermann, B. (2016). Bitcoin and beyond: A technical survey on decentralized digital currencies. *IEEE Communications Surveys and Tutorials*, *18*(3), 2084–2123. doi:10.1109/COMST.2016.2535718

Tseng, Y.-T., Jan, R.-H., Chen, C., Wang, C.-F., & Li, H.-H. (2010). A vehicle-density-based forwarding scheme for emergency message broadcasts in VANETs. In *The 7th IEEE International Conference on Mobile Ad-hoc and Sensor Systems (IEEE MASS 2010)* (pp. 703-708). IEEE.

Tuyisenge, L., Ayaida, M., Tohme, S., & Afilal, L.-E. (2018). Network architectures in internet of vehicles (iov): Review, protocols analysis, challenges and issues. In *International Conference on Internet of Vehicles*. Springer. 10.1007/978-3-030-05081-8_1

Tyagi, A. K., Agarwal, D., & Sreenath, N. (2022, January). SecVT: Securing the Vehicles of Tomorrow using Blockchain Technology. In *2022 International Conference on Computer Communication and Informatics (ICCCI)* (pp. 1-6). IEEE. 10.1109/ICCCI54379.2022.9740965

Uddin, M. A., Stranieri, A., Gondal, I., & Balasubramanian, V. (2021). A survey on the adoption of blockchain in iot: Challenges and solutions. *Blockchain: Research and Applications*, *2*(2), 100006. doi:10.1016/j.bcra.2021.100006

UN. (2017). *World population prospects*. https://www.un.org/development/desa/en/news/population/world-population-prospects-2017.html

Unmanned Cargo. (n.d.). *Drones going postal*. http://unmannedcargo.org/drones-going-postal-summary-postal-servicedelivery- drone-trials/

ur Rehman, Khan, Zia, & Zheng. (2013). Vehicular Ad -Hoc Networks (VANETs) – An Overview and Challenges. *JWNC*.

Ur Rehman, S., Khan, M. A., Zia, T. A., & Zheng, L. (2013). Vehicular ad-hoc networks (VANETs)-an overview and challenges. *Journal of Wireless Networking and Communications*, 29-38.

Van der Bergh, B., Chiumento, A., & Pollin, S. (2016). LTE in the sky: Trading off propagation benefits with interference costs for aerial nodes. *IEEE Communications Magazine*, *54*(5), 44–50. doi:10.1109/MCOM.2016.7470934

Vanamoorthy, M., & Chinnaiah, V. (2020). Congestion-free transient plane (CFTP) using bandwidth sharing during link failures in SDN. *The Computer Journal*, *63*(6), 832–843. doi:10.1093/comjnl/bxz137

Vanamoorthy, M., Chinnaiah, V., & Sekar, H. (2020). A hybrid approach for providing improved link connectivity in SDN. *The International Arab Journal of Information Technology*, *17*(2), 250–256. doi:10.34028/iajit/17/2/13

Vanitha, N., & Padmavathi, G. (2018, March). A comparative study on communication architecture of unmanned aerial vehicles and security analysis of false data dissemination attacks. In *2018 International Conference on Current Trends towards Converging Technologies (ICCTCT)* (pp. 1-8). IEEE. 10.1109/ICCTCT.2018.8550873

Vasilyev, G. S., Surzhik, D. I., Kuzichkin, O. R., & Kurilov, I. A. (2020). Algorithms for Adapting Communication Protocols of Fanet Networks. *Journal of Software*, *15*(4), 114–122. doi:10.17706/jsw.15.4.114-122

Vega, F. A., Ramirez, F. C., Saiz, M. P., & Rosua, F. O. (2015). Multi-temporal imaging using an unmanned aerial vehicle for monitoring a sunflower crop. *Biosystems Engineering*, *132*, 19–27. doi:10.1016/j.biosystemseng.2015.01.008

Vishwakarma, L., & Das, D. (2022). SmartCoin: A novel incentive mechanism for vehicles in intelligent transportation system based on consortium blockchain. *Vehicular Communications*, *33*, 100429. doi:10.1016/j.vehcom.2021.100429

Vo, H. T., Kundu, A., & Mohania, M. K. (2018, March). Research Directions in Blockchain Data Management and Analytics. *Proceedings of EDBT, 2018*, 445–448.

Wang, F., & Zaniolo, C. (2004, November). XBiT: an XML-based bitemporal data model. In *International Conference on Conceptual Modeling (ER)* (pp. 810-824). Springer.

Wang, H., Ding, G., Gao, F., Chen, J., Wang, J., & Wang, L. (2018). Power control in UAV-supported ultra-dense networks: Communications, caching, and energy transfer. *IEEE Communications Magazine*, *56*(6), 28–34. doi:10.1109/MCOM.2018.1700431

Wang, J., Chen, H., & Sun, Z. (2020). Context-Aware Quantification for VANET Security: A Markov Chain-Based Scheme. *IEEE Access: Practical Innovations, Open Solutions*, *8*, 173618–173626. doi:10.1109/ACCESS.2020.3017557

Wang, J., Jiang, C., Han, Z., Ren, Y., Maunder, R. G., & Hanzo, L. (2017). Taking drones to the next level: Cooperative distributed unmanned-aerial-vehicular networks for small and mini drones. *IEEE Vehicular Technology Magazine*, *12*(3), 73–82. doi:10.1109/MVT.2016.2645481

Wang, L., Chen, F., & Yin, H. (2016). Detecting and tracking vehicles in traffic by unmanned aerial vehicles. *Automation in Construction*, *72*, 294–308. doi:10.1016/j.autcon.2016.05.008

Wang, R., Taleb, T., Jamalipour, A., & Sun, B. (2009). Protocols for reliable data transport in space internet. *IEEE Communications Surveys and Tutorials*, *11*(2), 21–32. doi:10.1109/SURV.2009.090203

Wang, X., Ning, Z., & Wang, L. (2018). Offloading in internet of vehicles: A fog-enabled real-time traffic management system. *IEEE Transactions on Industrial Informatics*, *14*(10), 4568–4578. doi:10.1109/TII.2018.2816590

Wang, X., Sun, H., Long, Y., Zheng, L., Liu, H., & Li, M. (2018). Development of visualization system for agricultural UAV crop growth information collection. *IFAC-PapersOnLine*, *51*(17), 631–636. doi:10.1016/j.ifacol.2018.08.126

Wang, Y., Sun, T., Rao, G., & Li, D. (2018). Formation tracking in sparse airborne networks. *IEEE Journal on Selected Areas in Communications*, *36*(9), 2000–2014. doi:10.1109/JSAC.2018.2864374

Wang, Z., Li, M., Khaleghi, A. M., Xu, D., Lobos, A., Vo, C., Lien, J.-M., Liu, J., & Son, Y. J. (2013). DDDAMS-based crowd control via UAVs and UGVs. *Procedia Computer Science*, *18*, 2028–2035. doi:10.1016/j.procs.2013.05.372

Wan, J., Liu, J., Shao, Z., Vasilakos, A. V., Imran, M., & Zhou, K. (2016). Mobile crowd sensing for traffic prediction in internet of vehicles. *Sensors (Basel)*, *16*(1), 88. doi:10.339016010088 PMID:26761013

Wikipedia. (2017). Delivery Drone. Accessed: Dec. 2017. Available: https://en.wikipedia.org/wiki/Delivery_drone

Windows#. (n.d.). *Installing - SUMO Documentation*. https://sumo.dlr.de/docs/Installing/index.html

Wu, H., Tao, X., Zhang, N., & Shen, X. (2018). Cooperative UAV cluster-assisted terrestrial cellular networks for ubiquitous coverage. *IEEE Journal on Selected Areas in Communications*, *36*(9), 2045–2058. doi:10.1109/JSAC.2018.2864418

Wu, S. H., Chen, C. M., & Chen, M. S. (2009). An asymmetric and asynchronous energy conservation protocol for vehicular networks. *IEEE Transactions on Mobile Computing*, *9*(1), 98–111.

Xu, Zhou, Cheng, Lyu, Shi, Chen, & Shen. (2017). Internet of vehicles in big data era. *IEEE/CAA Journal of Automatica Sinica, 5*(1), 19–35.

Xu, C., Zhang, C., & Xu, J. (2019, June). vchain: Enabling verifiable boolean range queries over blockchain databases. In *Proceedings of the 2019 International Conference on Management of Data* (pp. 141-158). 10.1145/3299869.3300083

Yang, F., Wang, S., Li, J., Liu, Z., & Sun, Q. (2014). An overview of internet of vehicles. *China Communications*, *11*(10), 1–15. doi:10.1109/CC.2014.6969789

Yang, X., Mao, Y., Xu, Q., & Wang, L. (2022). Priority-Based Hybrid MAC Protocol for VANET with UAV-Enabled Roadside Units. *Wireless Communications and Mobile Computing*, *2022*, 1–13. doi:10.1155/2022/8697248

Yarinezhad, R., & Sarabi, A. (2019). A new routing algorithm for vehicular ad-hoc networks based on glowworm swarm optimization algorithm. *Journal of Artificial Intelligence and Data Mining*, *7*(1), 69–76.

Yassein, M. B., & Alhuda, N. (2016). Flying ad-hoc networks: Routing protocols, mobility models, issues. *International Journal of Advanced Computer Science and Applications*.

Yuan, C., Liu, Z., & Zhang, Y. (2015). UAV-based forest fire detection and tracking using image processing techniques. In *International Conference on Unmanned Aircraft Systems (ICUAS)* (pp. 639--643). IEEE. 10.1109/ICUAS.2015.7152345

Zaidi, S., Atiquzzaman, M., & Calafate, C. T. (2021). Internet of flying things (IoFT): A survey. *Computer Communications*, *165*, 53–74. doi:10.1016/j.comcom.2020.10.023

Zeadally, S., Hunt, R., Chen, Y. S., Irwin, A., & Hassan, A. (2012). Vehicular ad hoc networks (VANETS): Status, results, and challenges. *Telecommunication Systems*, *50*(4), 217–241. doi:10.100711235-010-9400-5

Zeng, Y., Zhang, R., & Lim, T. J. (2016). Wireless communications with unmanned aerial vehicles: Opportunities and challenges. *IEEE Communications Magazine*, *54*(5), 36–42. doi:10.1109/MCOM.2016.7470933

Zhang, D., Jiang, N., Li, H., & Wu, C. (2017, December). IOP Conference Series: Materials Science and Engineering. In *International Conference on Robotics andMechantronics (ICRoM2017)* (*Vol. 12*, p. 14). Academic Press.

Zhang, P., & Boulos, M. N. K. (2020). Blockchain solutions for healthcare. In Precision Medicine for Investigators, Practitioners and Providers (pp. 519-524). Academic Press. doi:10.1016/B978-0-12-819178-1.00050-2

Zhang, J., Chen, T., Zhong, S., Wang, J., Zhang, W., & Zuo, X. (2019). Aeronautical Ad Hoc Networking for the Internet-Above-the-Clouds. *Proceedings of the IEEE*, 868-911. 10.1109/JPROC.2019.2909694

Zhang, M., Ali, G. M. N., Chong, P. H. J., Seet, B. C., & Kumar, A. (2019). A novel hybrid mac protocol for basic safety message broadcasting in vehicular networks. *IEEE Transactions on Intelligent Transportation Systems*, *21*(10), 4269–4282.

Zhang, M., Chong, P. H. J., & Seet, B. C. (2019). Performance analysis and boost for a MAC protocol in vehicular networks. *IEEE Transactions on Vehicular Technology*, *68*(9), 8721–8728.

Zhang, W., Zhang, Z., & Chao, H.-C. (2017). Cooperative fog computing for dealing with big data in the internet of vehicles: Architecture and hierarchical resource management. *IEEE Communications Magazine*, *55*(12), 60–67. doi:10.1109/MCOM.2017.1700208

Zhang, Y., Xu, C., Cheng, N., Li, H., Yang, H., & Shen, X. (2019a). Chronos$^{+}$: An Accurate Blockchain-Based Time-Stamping Scheme for Cloud Storage. *IEEE Transactions on Services Computing*, *13*(2), 216–229. doi:10.1109/TSC.2019.2947476

Zhang, Y., Xu, C., Li, H., Yang, H., & Shen, X. S. (2019b). Chronos: Secure and accurate time-stamping scheme for digital files via blockchain. In *2019 IEEE International Conference on Communications (ICC)* (pp. 1-6). IEEE. 10.1109/ICC.2019.8762071

Zhang, Y., Zhao, J., & Cao, G. (2010). Roadcast: A popularity aware content sharing scheme in vanets. *Mobile Computing and Communications Review*, *13*(4), 1–14. doi:10.1145/1740437.1740439

Zhao, H., Wang, H., Wu, W., & Wei, J. (2018). Deployment algorithms for UAV airborne networks toward on-demand coverage. *IEEE Journal on Selected Areas in Communications*, *36*(9), 2015–2031. doi:10.1109/JSAC.2018.2864376

Zhao, J., He, C., Peng, C., & Zhang, X. (2022). Blockchain for effective renewable energy management in the intelligent transportation system. *Journal of Interconnection Networks*, *2141009*. Advance online publication. doi:10.1142/S0219265921410097

Zheng, Z., Sangaiah, A. K., & Wang, T. (2018). Adaptive communication protocols in flying ad hoc network. *IEEE Communications Magazine*, *56*(1), 136–142. doi:10.1109/MCOM.2017.1700323

Zheng, Z., Xie, S., Dai, H. N., Chen, X., & Wang, H. (2018). Blockchain challenges and opportunities: A survey. *International Journal of Web and Grid Services*, *14*(4), 352–375. doi:10.1504/IJWGS.2018.095647

Zhou, Z., Gao, C., Xu, C., Zhang, Y., Mumtaz, S., & Rodriguez, J. (2017). Social big-data-based content dissemination in internet of vehicles. *IEEE Transactions on Industrial Informatics*, *14*(2), 768–777. doi:10.1109/TII.2017.2733001

# Related References

To continue our tradition of advancing information science and technology research, we have compiled a list of recommended IGI Global readings. These references will provide additional information and guidance to further enrich your knowledge and assist you with your own research and future publications.

Abir, J. I., & Shamim, T. F. (2020). What Compels Journalists to Take a Step Back?: Contextualizing the Media Laws and Policies of Bangladesh. In S. Jamil (Ed.), *Handbook of Research on Combating Threats to Media Freedom and Journalist Safety* (pp. 38–53). IGI Global. https://doi.org/10.4018/978-1-7998-1298-2.ch003

Adesina, K., Ganiu, O., & R., O. S. (2018). Television as Vehicle for Community Development: A Study of Lotunlotun Programme on (B.C.O.S.) Television, Nigeria. In A. Salawu, & T. Owolabi (Eds.), *Exploring Journalism Practice and Perception in Developing Countries* (pp. 60-84). Hershey, PA: IGI Global. https://doi.org/doi:10.4018/978-1-5225-3376-4.ch004

Aggarwal, K., Singh, S. K., Chopra, M., & Kumar, S. (2022). Role of Social Media in the COVID-19 Pandemic: A Literature Review. In B. Gupta, D. Peraković, A. Abd El-Latif, & D. Gupta (Eds.), *Data Mining Approaches for Big Data and Sentiment Analysis in Social Media* (pp. 91–115). IGI Global. https://doi.org/10.4018/978-1-7998-8413-2.ch004

Ahmad, R. H., & Pathan, A. K. (2017). A Study on M2M (Machine to Machine) System and Communication: Its Security, Threats, and Intrusion Detection System. In M. Ferrag & A. Ahmim (Eds.), *Security Solutions and Applied Cryptography in Smart Grid Communications* (pp. 179–214). Hershey, PA: IGI Global. doi:10.4018/978-1-5225-1829-7.ch010

Akanni, T. M. (2018). In Search of Women-Supportive Media for Sustainable Development in Nigeria. In A. Salawu & T. Owolabi (Eds.), *Exploring Journalism Practice and Perception in Developing Countries* (pp. 126–149). Hershey, PA: IGI Global. doi:10.4018/978-1-5225-3376-4.ch007

Akçay, D. (2017). The Role of Social Media in Shaping Marketing Strategies in the Airline Industry. In V. Benson, R. Tuninga, & G. Saridakis (Eds.), *Analyzing the Strategic Role of Social Networking in Firm Growth and Productivity* (pp. 214–233). Hershey, PA: IGI Global. doi:10.4018/978-1-5225-0559-4.ch012

Akmese, Z. (2020). Media Literacy and Framing of Media Content. In N. Taskiran (Ed.), *Handbook of Research on Multidisciplinary Approaches to Literacy in the Digital Age* (pp. 73–87). IGI Global. doi:10.4018/978-1-7998-1534-1.ch005

Al-Jenaibi, B. (2021). Paradigms of Public Relations in an Age of Digitalization: Social Media Analytics in the UAE. In O. Yildiz (Ed.), *Recent Developments in Individual and Organizational Adoption of ICTs* (pp. 262–277). IGI Global. https://doi.org/10.4018/978-1-7998-3045-0.ch016

Al-Rabayah, W. A. (2017). Social Media as Social Customer Relationship Management Tool: Case of Jordan Medical Directory. In W. Al-Rabayah, R. Khasawneh, R. Abu-shamaa, & I. Alsmadi (Eds.), *Strategic Uses of Social Media for Improved Customer Retention* (pp. 108–123). Hershey, PA: IGI Global. doi:10.4018/978-1-5225-1686-6.ch006

Algül, A., & Akpınar, M. E. (2022). Hate Speech on Social Media: "Dunyaerkeklergunu" Hashtag on Twitter. In E. Öngün, N. Pembecioğlu, & U. Gündüz (Eds.), *Handbook of Research on Digital Citizenship and Management During Crises* (pp. 293–305). IGI Global. https://doi.org/10.4018/978-1-7998-8421-7.ch016

Almjeld, J. (2017). Getting "Girly" Online: The Case for Gendering Online Spaces. In E. Monske & K. Blair (Eds.), *Handbook of Research on Writing and Composing in the Age of MOOCs* (pp. 87–105). Hershey, PA: IGI Global. doi:10.4018/978-1-5225-1718-4.ch006

Alsalmi, J. M., & Shehata, A. M. (2022). Official Uses of Social Media. In M. Al-Suqri, O. Al-Shaqsi, & J. Alsalmi (Eds.), *Mass Communications and the Influence of Information During Times of Crises* (pp. 123–140). IGI Global. https://doi.org/10.4018/978-1-7998-7503-1.ch006

Altaş, A. (2017). Space as a Character in Narrative Advertising: A Qualitative Research on Country Promotion Works. In R. Yılmaz (Ed.), *Narrative Advertising Models and Conceptualization in the Digital Age* (pp. 303–319). Hershey, PA: IGI Global. doi:10.4018/978-1-5225-2373-4.ch017

Altıparmak, B. (2017). The Structural Transformation of Space in Turkish Television Commercials as a Narrative Component. In R. Yılmaz (Ed.), *Narrative Advertising Models and Conceptualization in the Digital Age* (pp. 153–166). Hershey, PA: IGI Global. doi:10.4018/978-1-5225-2373-4.ch009

Arda, Ö., & Akmeşe, Z. (2021). Media Ethics: Evaluation of Television News in the Context of the Media and Ethics Relationship. In M. Taskiran & F. Pinarbaşi (Eds.), *Multidisciplinary Approaches to Ethics in the Digital Era* (pp. 96–110). IGI Global. https://doi.org/10.4018/978-1-7998-4117-3.ch007

Arık, E. (2019). Popular Culture and Media Intellectuals: Relationship Between Popular Culture and Capitalism – The Characteristics of the Media Intellectuals. In O. Ozgen (Ed.), *Handbook of Research on Consumption, Media, and Popular Culture in the Global Age* (pp. 1–10). IGI Global. https://doi.org/10.4018/978-1-5225-8491-9.ch001

Aslan, F. (2021). Could There Be an Alternative Method of Media Literacy in Promoting Health in Children and Adolescents? Media Literacy and Health Promotion. In G. Sarı (Eds.), *Handbook of Research on Representing Health and Medicine in Modern Media* (pp. 191-199). IGI Global. https://doi.org/10.4018/978-1-7998-6825-5.ch013

Assay, B. E. (2018). Regulatory Compliance, Ethical Behaviour, and Sustainable Growth in Nigeria's Telecommunications Industry. In I. Oncioiu (Ed.), *Ethics and Decision-Making for Sustainable Business Practices* (pp. 90–108). Hershey, PA: IGI Global. doi:10.4018/978-1-5225-3773-1.ch006

Assensoh-Kodua, A. (2022). This Thing of Social Media!: Indeed a Platform for Running or Developing Business in the Financial Sector. In M. Ertz (Ed.), *Handbook of Research on the Platform Economy and the Evolution of E-Commerce* (pp. 389–414). IGI Global. https://doi.org/10.4018/978-1-7998-7545-1.ch017

Atar, Ö. G. (2019). Digital Media Literacy: In-Depth Interview With the Parents of the Students Who Use Digital Media. In G. Sarı (Eds.), *Handbook of Research on Children's Consumption of Digital Media* (pp. 139-155). IGI Global. https://doi.org/10.4018/978-1-5225-5733-3.ch011

Attié, E. A., Bouvet, A., & Guibert, J. (2022). The Stakes of Social Media: Analyzing User Sentiments. In B. Gupta, D. Peraković, A. Abd El-Latif, & D. Gupta (Eds.), *Data Mining Approaches for Big Data and Sentiment Analysis in Social Media* (pp. 196–222). IGI Global. https://doi.org/10.4018/978-1-7998-8413-2.ch009

Averweg, U. R., & Leaning, M. (2018). The Qualities and Potential of Social Media. In M. Khosrow-Pour, D.B.A. (Ed.), Encyclopedia of Information Science and Technology, Fourth Edition (pp. 7106-7115). Hershey, PA: IGI Global. doi:10.4018/978-1-5225-2255-3.ch617

Baarda, R. (2017). Digital Democracy in Authoritarian Russia: Opportunity for Participation, or Site of Kremlin Control? In R. Luppicini & R. Baarda (Eds.), *Digital Media Integration for Participatory Democracy* (pp. 87–100). Hershey, PA: IGI Global. doi:10.4018/978-1-5225-2463-2.ch005

Barbosa, C., & Pedro, L. (2019). Time Orientation and Media Use: The Rise of the Device and the Changing Nature of Our Time Perception. In L. Oliveira (Ed.), *Managing Screen Time in an Online Society* (pp. 78–98). IGI Global. https://doi.org/10.4018/978-1-5225-8163-5.ch004

Başal, B. (2017). Actor Effect: A Study on Historical Figures Who Have Shaped the Advertising Narration. In R. Yılmaz (Ed.), *Narrative Advertising Models and Conceptualization in the Digital Age* (pp. 34–60). Hershey, PA: IGI Global. doi:10.4018/978-1-5225-2373-4.ch003

Behjati, M., & Cosmas, J. (2017). Self-Organizing Network Solutions: A Principal Step Towards Real 4G and Beyond. In D. Singh (Ed.), *Routing Protocols and Architectural Solutions for Optimal Wireless Networks and Security* (pp. 241–253). Hershey, PA: IGI Global. doi:10.4018/978-1-5225-2342-0.ch011

Bekafigo, M., & Pingley, A. C. (2017). Do Campaigns "Go Negative" on Twitter? In Y. Ibrahim (Ed.), *Politics, Protest, and Empowerment in Digital Spaces* (pp. 178–191). Hershey, PA: IGI Global. doi:10.4018/978-1-5225-1862-4.ch011

Bekman, M. (2022). Interaction of Internet Addiction with FoMO: The Role of Digital Media. In E. Öngün, N. Pembecioğlu, & U. Gündüz (Eds.), *Handbook of Research on Digital Citizenship and Management During Crises* (pp. 116–133). IGI Global. https://doi.org/10.4018/978-1-7998-8421-7.ch007

Bishop, J. (2017). Developing and Validating the "This Is Why We Can't Have Nice Things Scale": Optimising Political Online Communities for Internet Trolling. In Y. Ibrahim (Ed.), *Politics, Protest, and Empowerment in Digital Spaces* (pp. 153–177). Hershey, PA: IGI Global. doi:10.4018/978-1-5225-1862-4.ch010

Bitrus-Ojiambo, U. A., & King'ori, M. E. (2020). Media and Child Rights in Africa: Narrative Analysis of Child Rights in Kenyan Media. In O. Oyero (Ed.), *Media and Its Role in Protecting the Rights of Children in Africa* (pp. 125–148). IGI Global. https://doi.org/10.4018/978-1-7998-0329-4.ch007

Black, S. (2019). Diversity and Inclusion: How to Avoid Bias and Social Media Blunders. In J. Joe & E. Knight (Eds.), *Social Media for Communication and Instruction in Academic Libraries* (pp. 100–118). IGI Global. https://doi.org/10.4018/978-1-5225-8097-3.ch007

Bolat, N. (2017). The Functions of the Narrator in Digital Advertising. In R. Yılmaz (Ed.), *Narrative Advertising Models and Conceptualization in the Digital Age* (pp. 184–201). Hershey, PA: IGI Global. doi:10.4018/978-1-5225-2373-4.ch011

Brown, M. A. Sr. (2017). SNIP: High Touch Approach to Communication. In *Solutions for High-Touch Communications in a High-Tech World* (pp. 71–88). Hershey, PA: IGI Global. doi:10.4018/978-1-5225-1897-6.ch004

Brown, M. A. Sr. (2017). Comparing FTF and Online Communication Knowledge. In *Solutions for High-Touch Communications in a High-Tech World* (pp. 103–113). Hershey, PA: IGI Global. doi:10.4018/978-1-5225-1897-6.ch006

Brown, M. A. Sr. (2017). Where Do We Go from Here? In *Solutions for High-Touch Communications in a High-Tech World* (pp. 137–159). Hershey, PA: IGI Global. doi:10.4018/978-1-5225-1897-6.ch008

Brown, M. A. Sr. (2017). Bridging the Communication Gap. In *Solutions for High-Touch Communications in a High-Tech World* (pp. 1–22). Hershey, PA: IGI Global. doi:10.4018/978-1-5225-1897-6.ch001

Brown, M. A. Sr. (2017). Key Strategies for Communication. In *Solutions for High-Touch Communications in a High-Tech World* (pp. 179–202). Hershey, PA: IGI Global. doi:10.4018/978-1-5225-1897-6.ch010

Bryant, K. N. (2017). WordUp!: Student Responses to Social Media in the Technical Writing Classroom. In K. Bryant (Ed.), *Engaging 21st Century Writers with Social Media* (pp. 231–245). Hershey, PA: IGI Global. doi:10.4018/978-1-5225-0562-4.ch014

Buck, E. H. (2017). Slacktivism, Supervision, and #Selfies: Illuminating Social Media Composition through Reception Theory. In K. Bryant (Ed.), *Engaging 21st Century Writers with Social Media* (pp. 163–178). Hershey, PA: IGI Global. doi:10.4018/978-1-5225-0562-4.ch010

Bull, R., & Pianosi, M. (2017). Social Media, Participation, and Citizenship: New Strategic Directions. In V. Benson, R. Tuninga, & G. Saridakis (Eds.), *Analyzing the Strategic Role of Social Networking in Firm Growth and Productivity* (pp. 76–94). Hershey, PA: IGI Global. doi:10.4018/978-1-5225-0559-4.ch005

Caldarola, G., D'Editâ, A., Falcone, A., Lo Blundo, M., & Mancini, M. (2020). Communicating Archaeology in a Social World: Social Media, Blogs, Websites, and Best Practices. In E. Proietti (Ed.), *Developing Effective Communication Skills in Archaeology* (pp. 259–284). IGI Global. https://doi.org/10.4018/978-1-7998-1059-9.ch013

Carbajal, D., & Ramirez, Q. A. (2022). Applying Theoretical Perspectives to Social Media Influencers: A Content Analysis on Social Media Influencers the LaBrant Family. In M. Al-Suqri, O. Al-Shaqsi, & J. Alsalmi (Eds.), *Mass Communications and the Influence of Information During Times of Crises* (pp. 69–98). IGI Global. https://doi.org/10.4018/978-1-7998-7503-1.ch004

Castellano, S., & Khelladi, I. (2017). Play It Like Beckham!: The Influence of Social Networks on E-Reputation – The Case of Sportspeople and Their Online Fan Base. In A. Mesquita (Ed.), *Research Paradigms and Contemporary Perspectives on Human-Technology Interaction* (pp. 43–61). Hershey, PA: IGI Global. doi:10.4018/978-1-5225-1868-6.ch003

Chepken, C. K. (2020). Mobile-Based Social Media, What Is Cutting?: Mobile-Based Social Media: Extensive Study Findings. In S. Kır (Ed.), *New Media and Visual Communication in Social Networks* (pp. 113–135). IGI Global. https://doi.org/10.4018/978-1-7998-1041-4.ch007

Chugh, R., & Joshi, M. (2017). Challenges of Knowledge Management amidst Rapidly Evolving Tools of Social Media. In R. Chugh (Ed.), *Harnessing Social Media as a Knowledge Management Tool* (pp. 299–314). Hershey, PA: IGI Global. doi:10.4018/978-1-5225-0495-5.ch014

Cole, A. W., & Salek, T. A. (2017). Adopting a Parasocial Connection to Overcome Professional Kakoethos in Online Health Information. In M. Folk & S. Apostel (Eds.), *Establishing and Evaluating Digital Ethos and Online Credibility* (pp. 104–120). Hershey, PA: IGI Global. doi:10.4018/978-1-5225-1072-7.ch006

Cossiavelou, V. (2017). ACTA as Media Gatekeeping Factor: The EU Role as Global Negotiator. *International Journal of Interdisciplinary Telecommunications and Networking*, *9*(1), 26–37. doi:10.4018/IJITN.2017010103

Costanza, F. (2017). Social Media Marketing and Value Co-Creation: A System Dynamics Approach. In S. Rozenes & Y. Cohen (Eds.), *Handbook of Research on Strategic Alliances and Value Co-Creation in the Service Industry* (pp. 205–230). Hershey, PA: IGI Global. doi:10.4018/978-1-5225-2084-9.ch011

Cyrek, B. (2019). The User With a Thousand Faces: Campbell's "Monomyth" and Media Usage Practices. In J. Kreft, S. Kuczamer-Kłopotowska, & A. Kalinowska-Żeleźnik (Eds.), *Myth in Modern Media Management and Marketing* (pp. 50–68). IGI Global. https://doi.org/10.4018/978-1-5225-9100-9.ch003

Deniz, Ş. (2020). Is Somebody Spying on Us?: Social Media Users' Privacy Awareness. In S. Kır (Ed.), *New Media and Visual Communication in Social Networks* (pp. 156–172). IGI Global. https://doi.org/10.4018/978-1-7998-1041-4.ch009

Di Virgilio, F., & Antonelli, G. (2018). Consumer Behavior, Trust, and Electronic Word-of-Mouth Communication: Developing an Online Purchase Intention Model. In F. Di Virgilio (Ed.), *Social Media for Knowledge Management Applications in Modern Organizations* (pp. 58–80). Hershey, PA: IGI Global. doi:10.4018/978-1-5225-2897-5.ch003

Dolanbay, H. (2022). The Transformation of Literacy and Media Literacy. In C. Lane (Ed.), *Handbook of Research on Acquiring 21st Century Literacy Skills Through Game-Based Learning* (pp. 363–380). IGI Global. https://doi.org/10.4018/978-1-7998-7271-9.ch019

Dunn, R. A., & Herrmann, A. F. (2020). Comic Con Communion: Gender, Cosplay, and Media Fandom. In R. Dunn (Ed.), *Multidisciplinary Perspectives on Media Fandom* (pp. 37–52). IGI Global. https://doi.org/10.4018/978-1-7998-3323-9.ch003

DuQuette, J. L. (2017). Lessons from Cypris Chat: Revisiting Virtual Communities as Communities. In G. Panconesi & M. Guida (Eds.), *Handbook of Research on Collaborative Teaching Practice in Virtual Learning Environments* (pp. 299–316). Hershey, PA: IGI Global. doi:10.4018/978-1-5225-2426-7.ch016

Ekhlassi, A., Niknejhad Moghadam, M., & Adibi, A. (2018). The Concept of Social Media: The Functional Building Blocks. In *Building Brand Identity in the Age of Social Media: Emerging Research and Opportunities* (pp. 29–60). Hershey, PA: IGI Global. doi:10.4018/978-1-5225-5143-0.ch002

Ekhlassi, A., Niknejhad Moghadam, M., & Adibi, A. (2018). Social Media Branding Strategy: Social Media Marketing Approach. In *Building Brand Identity in the Age of Social Media: Emerging Research and Opportunities* (pp. 94–117). Hershey, PA: IGI Global. doi:10.4018/978-1-5225-5143-0.ch004

Ekhlassi, A., Niknejhad Moghadam, M., & Adibi, A. (2018). The Impact of Social Media on Brand Loyalty: Achieving "E-Trust" Through Engagement. In *Building Brand Identity in the Age of Social Media: Emerging Research and Opportunities* (pp. 155–168). Hershey, PA: IGI Global. doi:10.4018/978-1-5225-5143-0.ch007

El-Henawy, W. M. (2019). Media Literacy in EFL Teacher Education: A Necessity for 21st Century English Language Instruction. In M. Yildiz, M. Fazal, M. Ahn, R. Feirsen, & S. Ozdemir (Eds.), *Handbook of Research on Media Literacy Research and Applications Across Disciplines* (pp. 65–89). IGI Global. https://doi.org/10.4018/978-1-5225-9261-7.ch005

Elegbe, O. (2017). An Assessment of Media Contribution to Behaviour Change and HIV Prevention in Nigeria. In O. Nelson, B. Ojebuyi, & A. Salawu (Eds.), *Impacts of the Media on African Socio-Economic Development* (pp. 261–280). Hershey, PA: IGI Global. doi:10.4018/978-1-5225-1859-4.ch017

Endong, F. P. (2018). Hashtag Activism and the Transnationalization of Nigerian-Born Movements Against Terrorism: A Critical Appraisal of the #BringBackOurGirls Campaign. In F. Endong (Ed.), *Exploring the Role of Social Media in Transnational Advocacy* (pp. 36–54). Hershey, PA: IGI Global. doi:10.4018/978-1-5225-2854-8.ch003

Erkek, S. (2021). Health Communication and Social Media: A Study About Using Social Media in Medicine Companies. In G. Sarı (Eds.), *Handbook of Research on Representing Health and Medicine in Modern Media* (pp. 70-83). IGI Global. https://doi.org/10.4018/978-1-7998-6825-5.ch005

Erragcha, N. (2017). Using Social Media Tools in Marketing: Opportunities and Challenges. In M. Brown Sr., (Ed.), *Social Media Performance Evaluation and Success Measurements* (pp. 106–129). Hershey, PA: IGI Global. doi:10.4018/978-1-5225-1963-8.ch006

Ersoy, M. (2019). Social Media and Children. In G. Sarı (Ed.), *Handbook of Research on Children's Consumption of Digital Media* (pp. 11-23). IGI Global. https://doi.org/10.4018/978-1-5225-5733-3.ch002

Ezeh, N. C. (2018). Media Campaign on Exclusive Breastfeeding: Awareness, Perception, and Acceptability Among Mothers in Anambra State, Nigeria. In A. Salawu & T. Owolabi (Eds.), *Exploring Journalism Practice and Perception in Developing Countries* (pp. 172–193). Hershey, PA: IGI Global. doi:10.4018/978-1-5225-3376-4.ch009

Fawole, O. A., & Osho, O. A. (2017). Influence of Social Media on Dating Relationships of Emerging Adults in Nigerian Universities: Social Media and Dating in Nigeria. In M. Wright (Ed.), *Identity, Sexuality, and Relationships among Emerging Adults in the Digital Age* (pp. 168–177). Hershey, PA: IGI Global. doi:10.4018/978-1-5225-1856-3.ch011

Fayoyin, A. (2017). Electoral Polling and Reporting in Africa: Professional and Policy Implications for Media Practice and Political Communication in a Digital Age. In N. Mhiripiri & T. Chari (Eds.), *Media Law, Ethics, and Policy in the Digital Age* (pp. 164–181). Hershey, PA: IGI Global. doi:10.4018/978-1-5225-2095-5.ch009

Fayoyin, A. (2018). Rethinking Media Engagement Strategies for Social Change in Africa: Context, Approaches, and Implications for Development Communication. In A. Salawu & T. Owolabi (Eds.), *Exploring Journalism Practice and Perception in Developing Countries* (pp. 257–280). Hershey, PA: IGI Global. doi:10.4018/978-1-5225-3376-4.ch013

Fechine, Y., & Rêgo, S. C. (2018). Transmedia Television Journalism in Brazil: Jornal da Record News as Reference. In R. Gambarato & G. Alzamora (Eds.), *Exploring Transmedia Journalism in the Digital Age* (pp. 253–265). Hershey, PA: IGI Global. doi:10.4018/978-1-5225-3781-6.ch015

Fener, E. (2021). Social Media and Health Communication. In G. Sarı (Ed.), *Handbook of Research on Representing Health and Medicine in Modern Media* (pp. 16-32). IGI Global. https://doi.org/10.4018/978-1-7998-6825-5.ch002

Fernandes dos Santos, N. (2020). The Use of Twitter During the 2013 Protests in Brazil: Mainstream Media at Stake. In A. Solo (Ed.), *Handbook of Research on Politics in the Computer Age* (pp. 181–202). IGI Global. https://doi.org/10.4018/978-1-7998-0377-5.ch011

Fiore, C. (2017). The Blogging Method: Improving Traditional Student Writing Practices. In K. Bryant (Ed.), *Engaging 21st Century Writers with Social Media* (pp. 179–198). Hershey, PA: IGI Global. doi:10.4018/978-1-5225-0562-4.ch011

Friesem, E., & Friesem, Y. (2019). Media Literacy Education in the Era of Post-Truth: Paradigm Crisis. In M. Yildiz, M. Fazal, M. Ahn, R. Feirsen, & S. Ozdemir (Eds.), *Handbook of Research on Media Literacy Research and Applications Across Disciplines* (pp. 119–134). IGI Global. https://doi.org/10.4018/978-1-5225-9261-7.ch008

Fung, Y., Lee, L., Chui, K. T., Cheung, G. H., Tang, C., & Wong, S. (2022). Sentiment Analysis and Summarization of Facebook Posts on News Media. In B. Gupta, D. Peraković, A. Abd El-Latif, & D. Gupta (Eds.), *Data Mining Approaches for Big Data and Sentiment Analysis in Social Media* (pp. 142–154). IGI Global. https://doi.org/10.4018/978-1-7998-8413-2.ch006

Gambarato, R. R., Alzamora, G. C., & Tárcia, L. P. (2018). 2016 Rio Summer Olympics and the Transmedia Journalism of Planned Events. In R. Gambarato & G. Alzamora (Eds.), *Exploring Transmedia Journalism in the Digital Age* (pp. 126–146). Hershey, PA: IGI Global. doi:10.4018/978-1-5225-3781-6.ch008

Ganguin, S., Gemkow, J., & Haubold, R. (2017). Information Overload as a Challenge and Changing Point for Educational Media Literacies. In R. Marques & J. Batista (Eds.), *Information and Communication Overload in the Digital Age* (pp. 302–328). Hershey, PA: IGI Global. doi:10.4018/978-1-5225-2061-0.ch013

Gardner, G. C. (2017). The Lived Experience of Smartphone Use in a Unit of the United States Army. In F. Topor (Ed.), *Handbook of Research on Individualism and Identity in the Globalized Digital Age* (pp. 88–117). Hershey, PA: IGI Global. doi:10.4018/978-1-5225-0522-8.ch005

Garg, P., & Pahuja, S. (2020). Social Media: Concept, Role, Categories, Trends, Social Media and AI, Impact on Youth, Careers, Recommendations. In S. Alavi & V. Ahuja (Eds.), *Managing Social Media Practices in the Digital Economy* (pp. 172–192). IGI Global. https://doi.org/10.4018/978-1-7998-2185-4.ch008

Golightly, D., & Houghton, R. J. (2018). Social Media as a Tool to Understand Behaviour on the Railways. In S. Kohli, A. Kumar, J. Easton, & C. Roberts (Eds.), *Innovative Applications of Big Data in the Railway Industry* (pp. 224–239). Hershey, PA: IGI Global. doi:10.4018/978-1-5225-3176-0.ch010

Gouveia, P. (2020). The New Media vs. Old Media Trap: How Contemporary Arts Became Playful Transmedia Environments. In C. Soares & E. Simão (Eds.), *Multidisciplinary Perspectives on New Media Art* (pp. 25–46). IGI Global. https://doi.org/10.4018/978-1-7998-3669-8.ch002

Gundogan, M. B. (2017). In Search for a "Good Fit" Between Augmented Reality and Mobile Learning Ecosystem. In G. Kurubacak & H. Altinpulluk (Eds.), *Mobile Technologies and Augmented Reality in Open Education* (pp. 135–153). Hershey, PA: IGI Global. doi:10.4018/978-1-5225-2110-5.ch007

Gupta, H. (2018). Impact of Digital Communication on Consumer Behaviour Processes in Luxury Branding Segment: A Study of Apparel Industry. In S. Dasgupta, S. Biswal, & M. Ramesh (Eds.), *Holistic Approaches to Brand Culture and Communication Across Industries* (pp. 132–157). Hershey, PA: IGI Global. doi:10.4018/978-1-5225-3150-0.ch008

Guzman-Garcia, P. A., Orozco-Quintana, E., Sepulveda-Gonzalez, D., Cooley-Magallanes, A., Salas-Velazquez, D., Lopez-Garcia, C., Ramírez-Treviño, A., Espinoza-Moran, A. L., Lopez, M., & Segura-Azuara, N. D. (2022). Ending Health Promotion Lethargy: A Social Media Awareness Campaign to Face Hypothyroidism. In M. Lopez (Ed.), *Advancing Health Education With Telemedicine* (pp. 165–182). IGI Global. https://doi.org/10.4018/978-1-7998-8783-6.ch009

Hafeez, E., & Zahid, L. (2021). Sexism and Gender Discrimination in Pakistan's Mainstream News Media. In S. Jamil, B. Çoban, B. Ataman, & G. Appiah-Adjei (Eds.), *Handbook of Research on Discrimination, Gender Disparity, and Safety Risks in Journalism* (pp. 60–89). IGI Global. https://doi.org/10.4018/978-1-7998-6686-2.ch005

Hai-Jew, S. (2017). Creating "(Social) Network Art" with NodeXL. In S. Hai-Jew (Ed.), *Social Media Data Extraction and Content Analysis* (pp. 342–393). Hershey, PA: IGI Global. doi:10.4018/978-1-5225-0648-5.ch011

Hai-Jew, S. (2017). Employing the Sentiment Analysis Tool in NVivo 11 Plus on Social Media Data: Eight Initial Case Types. In N. Rao (Ed.), *Social Media Listening and Monitoring for Business Applications* (pp. 175–244). Hershey, PA: IGI Global. doi:10.4018/978-1-5225-0846-5.ch010

Hai-Jew, S. (2017). Conducting Sentiment Analysis and Post-Sentiment Data Exploration through Automated Means. In S. Hai-Jew (Ed.), *Social Media Data Extraction and Content Analysis* (pp. 202–240). Hershey, PA: IGI Global. doi:10.4018/978-1-5225-0648-5.ch008

Hai-Jew, S. (2017). Applied Analytical "Distant Reading" using NVivo 11 Plus. In S. Hai-Jew (Ed.), *Social Media Data Extraction and Content Analysis* (pp. 159–201). Hershey, PA: IGI Global. doi:10.4018/978-1-5225-0648-5.ch007

Hai-Jew, S. (2017). Flickering Emotions: Feeling-Based Associations from Related Tags Networks on Flickr. In S. Hai-Jew (Ed.), *Social Media Data Extraction and Content Analysis* (pp. 296–341). Hershey, PA: IGI Global. doi:10.4018/978-1-5225-0648-5.ch010

Hai-Jew, S. (2017). Manually Profiling Egos and Entities across Social Media Platforms: Evaluating Shared Messaging and Contents, User Networks, and Metadata. In V. Benson, R. Tuninga, & G. Saridakis (Eds.), *Analyzing the Strategic Role of Social Networking in Firm Growth and Productivity* (pp. 352–405). Hershey, PA: IGI Global. doi:10.4018/978-1-5225-0559-4.ch019

Hai-Jew, S. (2017). Exploring "User," "Video," and (Pseudo) Multi-Mode Networks on YouTube with NodeXL. In S. Hai-Jew (Ed.), *Social Media Data Extraction and Content Analysis* (pp. 242–295). Hershey, PA: IGI Global. doi:10.4018/978-1-5225-0648-5.ch009

Hai-Jew, S. (2018). Exploring "Mass Surveillance" Through Computational Linguistic Analysis of Five Text Corpora: Academic, Mainstream Journalism, Microblogging Hashtag Conversation, Wikipedia Articles, and Leaked Government Data. In *Techniques for Coding Imagery and Multimedia: Emerging Research and Opportunities* (pp. 212–286). Hershey, PA: IGI Global. doi:10.4018/978-1-5225-2679-7.ch004

Hai-Jew, S. (2018). Exploring Identity-Based Humor in a #Selfies #Humor Image Set From Instagram. In *Techniques for Coding Imagery and Multimedia: Emerging Research and Opportunities* (pp. 1–90). Hershey, PA: IGI Global. doi:10.4018/978-1-5225-2679-7.ch001

Hai-Jew, S. (2018). See Ya!: Exploring American Renunciation of Citizenship Through Targeted and Sparse Social Media Data Sets and a Custom Spatial-Based Linguistic Analysis Dictionary. In *Techniques for Coding Imagery and Multimedia: Emerging Research and Opportunities* (pp. 287–393). Hershey, PA: IGI Global. doi:10.4018/978-1-5225-2679-7.ch005

Hasan, H., & Linger, H. (2017). Connected Living for Positive Ageing. In S. Gordon (Ed.), *Online Communities as Agents of Change and Social Movements* (pp. 203–223). Hershey, PA: IGI Global. doi:10.4018/978-1-5225-2495-3.ch008

Hersey, L. N. (2017). CHOICES: Measuring Return on Investment in a Nonprofit Organization. In M. Brown Sr., (Ed.), *Social Media Performance Evaluation and Success Measurements* (pp. 157–179). Hershey, PA: IGI Global. doi:10.4018/978-1-5225-1963-8.ch008

Heuva, W. E. (2017). Deferring Citizens' "Right to Know" in an Information Age: The Information Deficit in Namibia. In N. Mhiripiri & T. Chari (Eds.), *Media Law, Ethics, and Policy in the Digital Age* (pp. 245–267). Hershey, PA: IGI Global. doi:10.4018/978-1-5225-2095-5.ch014

Hopwood, M., & McLean, H. (2017). Social Media in Crisis Communication: The Lance Armstrong Saga. In V. Benson, R. Tuninga, & G. Saridakis (Eds.), *Analyzing the Strategic Role of Social Networking in Firm Growth and Productivity* (pp. 45–58). Hershey, PA: IGI Global. doi:10.4018/978-1-5225-0559-4.ch003

Horst, S., & Murschetz, P. C. (2019). Strategic Media Entrepreneurship: Theory Development and Problematization. *Journal of Media Management and Entrepreneurship, 1*(1), 1–26. https://doi.org/10.4018/JMME.2019010101

Hotur, S. K. (2018). Indian Approaches to E-Diplomacy: An Overview. In S. Bute (Ed.), *Media Diplomacy and Its Evolving Role in the Current Geopolitical Climate* (pp. 27–35). Hershey, PA: IGI Global. doi:10.4018/978-1-5225-3859-2.ch002

Inder, S. (2021). Social Media, Crowdsourcing, and Marketing. In A. Singh (Ed.), *Big Data Analytics for Improved Accuracy, Efficiency, and Decision Making in Digital Marketing* (pp. 64–73). IGI Global. https://doi.org/10.4018/978-1-7998-7231-3.ch005

Işık, T. (2021). Media and Health Communication Campaigns. In G. Sarı (Ed.), *Handbook of Research on Representing Health and Medicine in Modern Media* (pp. 1-15). IGI Global. https://doi.org/10.4018/978-1-7998-6825-5.ch001

Iwasaki, Y. (2017). Youth Engagement in the Era of New Media. In M. Adria & Y. Mao (Eds.), *Handbook of Research on Citizen Engagement and Public Participation in the Era of New Media* (pp. 90–105). Hershey, PA: IGI Global. doi:10.4018/978-1-5225-1081-9.ch006

Jamieson, H. V. (2017). We have a Situation!: Cyberformance and Civic Engagement in Post-Democracy. In R. Shin (Ed.), *Convergence of Contemporary Art, Visual Culture, and Global Civic Engagement* (pp. 297–317). Hershey, PA: IGI Global. doi:10.4018/978-1-5225-1665-1.ch017

Jimoh, J., & Kayode, J. (2018). Imperative of Peace and Conflict-Sensitive Journalism in Development. In A. Salawu & T. Owolabi (Eds.), *Exploring Journalism Practice and Perception in Developing Countries* (pp. 150–171). Hershey, PA: IGI Global. doi:10.4018/978-1-5225-3376-4.ch008

Joseph, J. J., & Florea, D. (2020). Clinical Topics in Social Media: The Role of Self-Disclosing on Social Media for Friendship and Identity in Specialized Populations. In M. Desjarlais (Ed.), *The Psychology and Dynamics Behind Social Media Interactions* (pp. 28–56). IGI Global. https://doi.org/10.4018/978-1-5225-9412-3.ch002

Kaale, K. B., & Mgeta, M. B. (2020). Photojournalism Ethics: Portraying Children's Photos in Tanzanian Media. In O. Oyero (Ed.), *Media and Its Role in Protecting the Rights of Children in Africa* (pp. 149–168). IGI Global. https://doi.org/10.4018/978-1-7998-0329-4.ch008

Kanellopoulos, D. N. (2018). Group Synchronization for Multimedia Systems. In M. Khosrow-Pour, D.B.A. (Ed.), Encyclopedia of Information Science and Technology, Fourth Edition (pp. 6435-6446). Hershey, PA: IGI Global. doi:10.4018/978-1-5225-2255-3.ch559

Kapepo, M. I., & Mayisela, T. (2017). Integrating Digital Literacies Into an Undergraduate Course: Inclusiveness Through Use of ICTs. In C. Ayo & V. Mbarika (Eds.), *Sustainable ICT Adoption and Integration for Socio-Economic Development* (pp. 152–173). Hershey, PA: IGI Global. doi:10.4018/978-1-5225-2565-3.ch007

Karahoca, A., & Yengin, İ. (2018). Understanding the Potentials of Social Media in Collaborative Learning. In M. Khosrow-Pour, D.B.A. (Ed.), Encyclopedia of Information Science and Technology, Fourth Edition (pp. 7168-7180). Hershey, PA: IGI Global. doi:10.4018/978-1-5225-2255-3.ch623

Kasemsap, K. (2017). Professional and Business Applications of Social Media Platforms. In V. Benson, R. Tuninga, & G. Saridakis (Eds.), *Analyzing the Strategic Role of Social Networking in Firm Growth and Productivity* (pp. 427–450). Hershey, PA: IGI Global. doi:10.4018/978-1-5225-0559-4.ch021

Kasemsap, K. (2017). Mastering Social Media in the Modern Business World. In N. Rao (Ed.), *Social Media Listening and Monitoring for Business Applications* (pp. 18–44). Hershey, PA: IGI Global. doi:10.4018/978-1-5225-0846-5.ch002

Kaufmann, H. R., & Manarioti, A. (2017). Consumer Engagement in Social Media Platforms. In *Encouraging Participative Consumerism Through Evolutionary Digital Marketing: Emerging Research and Opportunities* (pp. 95–123). Hershey, PA: IGI Global. doi:10.4018/978-1-68318-012-8.ch004

Kavak, B., Özdemir, N., & Erol-Boyacı, G. (2020). A Literature Review of Social Media for Marketing: Social Media Use in B2C and B2B Contexts. In S. Alavi & V. Ahuja (Eds.), *Managing Social Media Practices in the Digital Economy* (pp. 67–96). IGI Global. https://doi.org/10.4018/978-1-7998-2185-4.ch004

Kavoura, A., & Kefallonitis, E. (2018). The Effect of Social Media Networking in the Travel Industry. In M. Khosrow-Pour, D.B.A. (Ed.), Encyclopedia of Information Science and Technology, Fourth Edition (pp. 4052-4063). Hershey, PA: IGI Global. doi:10.4018/978-1-5225-2255-3.ch351

Kawamura, Y. (2018). Practice and Modeling of Advertising Communication Strategy: Sender-Driven and Receiver-Driven. In T. Ogata & S. Asakawa (Eds.), *Content Generation Through Narrative Communication and Simulation* (pp. 358–379). Hershey, PA: IGI Global. doi:10.4018/978-1-5225-4775-4.ch013

Kaya, A., & Mantar, O. B. (2021). Social Media and Health Communication: Vaccine Refusal/Hesitancy. In G. Sarı (Ed.), *Handbook of Research on Representing Health and Medicine in Modern Media* (pp. 33-53). IGI Global. https://doi.org/10.4018/978-1-7998-6825-5.ch003

Kaya, A. Y., & Ata, F. (2022). New Media and Digital Paranoia: Extreme Skepticism in Digital Communication. In H. Aker & M. Aiken (Eds.), *Handbook of Research on Cyberchondria, Health Literacy, and the Role of Media in Society's Perception of Medical Information* (pp. 330–343). IGI Global. https://doi.org/10.4018/978-1-7998-8630-3.ch018

Kell, C., & Czerniewicz, L. (2017). Visibility of Scholarly Research and Changing Research Communication Practices: A Case Study from Namibia. In A. Esposito (Ed.), *Research 2.0 and the Impact of Digital Technologies on Scholarly Inquiry* (pp. 97–116). Hershey, PA: IGI Global. doi:10.4018/978-1-5225-0830-4.ch006

Kharade, S. S. (2022). An Adverse Effect of Social, Gaming, and Entertainment Media on Overall Development of Adolescents. In S. Malik, R. Bansal, & A. Tyagi (Eds.), *Impact and Role of Digital Technologies in Adolescent Lives* (pp. 26–34). IGI Global. https://doi.org/10.4018/978-1-7998-8318-0.ch003

Kılınç, U. (2017). Create It! Extend It!: Evolution of Comics Through Narrative Advertising. In R. Yılmaz (Ed.), *Narrative Advertising Models and Conceptualization in the Digital Age* (pp. 117–132). Hershey, PA: IGI Global. doi:10.4018/978-1-5225-2373-4.ch007

Kocakoç, I. D., & Özkan, P. (2022). Clubhouse Experience: Sentiment Analysis of an Alternative Platform From the Eyes of Classic Social Media Users. In B. Gupta, D. Peraković, A. Abd El-Latif, & D. Gupta (Eds.), *Data Mining Approaches for Big Data and Sentiment Analysis in Social Media* (pp. 244–264). IGI Global. https://doi.org/10.4018/978-1-7998-8413-2.ch011

Kreft, J. (2019). A Myth and Media Management: The Facade Rhetoric and Business Objectives. In J. Kreft, S. Kuczamer-Kłopotowska, & A. Kalinowska-Żeleźnik (Eds.), *Myth in Modern Media Management and Marketing* (pp. 118–141). IGI Global. https://doi.org/10.4018/978-1-5225-9100-9.ch006

Krishnamurthy, R. (2019). Social Media as a Marketing Tool. In P. Mishra & S. Dham (Eds.), *Application of Gaming in New Media Marketing* (pp. 181–201). IGI Global. https://doi.org/10.4018/978-1-5225-6064-7.ch011

Kumar, D., & Gupta, P. (2021). Communicating in Media Dark Areas. In R. Jackson & A. Reboulet (Eds.), *Effective Strategies for Communicating Insights in Business* (pp. 141-156). IGI Global. https://doi.org/10.4018/978-1-7998-3964-4.ch009

Kumar, P., & Sinha, A. (2018). Business-Oriented Analytics With Social Network of Things. In H. Bansal, G. Shrivastava, G. Nguyen, & L. Stanciu (Eds.), *Social Network Analytics for Contemporary Business Organizations* (pp. 166–187). Hershey, PA: IGI Global. doi:10.4018/978-1-5225-5097-6.ch009

Kunock, A. I. (2017). Boko Haram Insurgency in Cameroon: Role of Mass Media in Conflict Management. In N. Mhiripiri & T. Chari (Eds.), *Media Law, Ethics, and Policy in the Digital Age* (pp. 226–244). Hershey, PA: IGI Global. doi:10.4018/978-1-5225-2095-5.ch013

Labadie, J. A. (2018). Digitally Mediated Art Inspired by Technology Integration: A Personal Journey. In A. Ursyn (Ed.), *Visual Approaches to Cognitive Education With Technology Integration* (pp. 121–162). Hershey, PA: IGI Global. doi:10.4018/978-1-5225-5332-8.ch008

Lantz, E. (2020). Immersion Domes: Next-Generation Arts and Entertainment Venues. In J. Morie & K. McCallum (Eds.), *Handbook of Research on the Global Impacts and Roles of Immersive Media* (pp. 314–346). IGI Global. https://doi.org/10.4018/978-1-7998-2433-6.ch016

Lasisi, M. I., Adebiyi, R. A., & Ajetunmobi, U. O. (2020). Predicting Migration to Developed Countries: The Place of Media Attention. In N. Okorie, B. Ojebuyi, & J. Macharia (Eds.), *Handbook of Research on the Global Impact of Media on Migration Issues* (pp. 293–311). IGI Global. https://doi.org/10.4018/978-1-7998-0210-5.ch017

Lefkowith, S. (2017). Credibility and Crisis in Pseudonymous Communities. In M. Folk & S. Apostel (Eds.), *Establishing and Evaluating Digital Ethos and Online Credibility* (pp. 190–236). Hershey, PA: IGI Global. doi:10.4018/978-1-5225-1072-7.ch010

Lekic-Subasic, Z. (2021). Women and Media: What Public Service Media Can Do to Ensure Gender Equality. In S. Jamil, B. Çoban, B. Ataman, & G. Appiah-Adjei (Eds.), *Handbook of Research on Discrimination, Gender Disparity, and Safety Risks in Journalism* (pp. 8–23). IGI Global. https://doi.org/10.4018/978-1-7998-6686-2.ch002

Luppicini, R. (2017). Technoethics and Digital Democracy for Future Citizens. In R. Luppicini & R. Baarda (Eds.), *Digital Media Integration for Participatory Democracy* (pp. 1–21). Hershey, PA: IGI Global. doi:10.4018/978-1-5225-2463-2.ch001

Maher, D. (2018). Supporting Pre-Service Teachers' Understanding and Use of Mobile Devices. In J. Keengwe (Ed.), *Handbook of Research on Mobile Technology, Constructivism, and Meaningful Learning* (pp. 160–177). Hershey, PA: IGI Global. doi:10.4018/978-1-5225-3949-0.ch009

Makhwanya, A. (2018). Barriers to Social Media Advocacy: Lessons Learnt From the Project "Tell Them We Are From Here". In F. Endong (Ed.), *Exploring the Role of Social Media in Transnational Advocacy* (pp. 55–72). Hershey, PA: IGI Global. doi:10.4018/978-1-5225-2854-8.ch004

Malicki-Sanchez, K. (2020). Out of Our Minds: Ontology and Embodied Media in a Post-Human Paradigm. In J. Morie & K. McCallum (Eds.), *Handbook of Research on the Global Impacts and Roles of Immersive Media* (pp. 10–36). IGI Global. https://doi.org/10.4018/978-1-7998-2433-6.ch002

Manli, G., & Rezaei, S. (2017). Value and Risk: Dual Pillars of Apps Usefulness. In S. Rezaei (Ed.), *Apps Management and E-Commerce Transactions in Real-Time* (pp. 274–292). Hershey, PA: IGI Global. doi:10.4018/978-1-5225-2449-6.ch013

Manrique, C. G., & Manrique, G. G. (2017). Social Media's Role in Alleviating Political Corruption and Scandals: The Philippines during and after the Marcos Regime. In K. Demirhan & D. Çakır-Demirhan (Eds.), *Political Scandal, Corruption, and Legitimacy in the Age of Social Media* (pp. 205–222). Hershey, PA: IGI Global. doi:10.4018/978-1-5225-2019-1.ch009

Marjerison, R. K., Lin, Y., & Kennedyd, S. I. (2019). An Examination of Motivation and Media Type: Sharing Content on Chinese Social Media. *International Journal of Social Media and Online Communities*, *11*(1), 15–34. https://doi.org/10.4018/IJSMOC.2019010102

Marovitz, M. (2017). Social Networking Engagement and Crisis Communication Considerations. In M. Brown Sr., (Ed.), *Social Media Performance Evaluation and Success Measurements* (pp. 130–155). Hershey, PA: IGI Global. doi:10.4018/978-1-5225-1963-8.ch007

Martin, P. M., & Onampally, J. J. (2019). Patterns of Deceptive Communication of Social and Religious Issues in Social Media: Representation of Social Issues in Social Media. In I. Chiluwa & S. Samoilenko (Eds.), *Handbook of Research on Deception, Fake News, and Misinformation Online* (pp. 490–502). IGI Global. https://doi.org/10.4018/978-1-5225-8535-0.ch026

Masterson, J. R. (2020). Chinese Citizenry Social Media Pressures and Public Official Responses: The Double-Edged Sword of Social Media in China. In S. Edwards III & D. Santos (Eds.), *Digital Transformation and Its Role in Progressing the Relationship Between States and Their Citizens* (pp. 139-181). IGI Global. https://doi.org/10.4018/978-1-7998-3152-5.ch007

Maulana, I. (2018). Spontaneous Taking and Posting Selfie: Reclaiming the Lost Trust. In S. Hai-Jew (Ed.), *Selfies as a Mode of Social Media and Work Space Research* (pp. 28–50). Hershey, PA: IGI Global. doi:10.4018/978-1-5225-3373-3.ch002

Mayo, S. (2018). A Collective Consciousness Model in a Post-Media Society. In M. Khosrow-Pour (Ed.), *Enhancing Art, Culture, and Design With Technological Integration* (pp. 25–49). Hershey, PA: IGI Global. doi:10.4018/978-1-5225-5023-5.ch002

Mazur, E., Signorella, M. L., & Hough, M. (2018). The Internet Behavior of Older Adults. In M. Khosrow-Pour, D.B.A. (Ed.), Encyclopedia of Information Science and Technology, Fourth Edition (pp. 7026-7035). Hershey, PA: IGI Global. doi:10.4018/978-1-5225-2255-3.ch609

McCallum, K. M. (2020). Immersive Experience: Convergence, Storyworlds, and the Power for Social Impact. In J. Morie & K. McCallum (Eds.), *Handbook of Research on the Global Impacts and Roles of Immersive Media* (pp. 453–484). IGI Global. https://doi.org/10.4018/978-1-7998-2433-6.ch022

McGuire, M. (2017). Reblogging as Writing: The Role of Tumblr in the Writing Classroom. In K. Bryant (Ed.), *Engaging 21st Century Writers with Social Media* (pp. 116–131). Hershey, PA: IGI Global. doi:10.4018/978-1-5225-0562-4.ch007

McKee, J. (2018). Architecture as a Tool to Solve Business Planning Problems. In M. Khosrow-Pour, D.B.A. (Ed.), Encyclopedia of Information Science and Technology, Fourth Edition (pp. 573-586). Hershey, PA: IGI Global. doi:10.4018/978-1-5225-2255-3.ch050

McMahon, D. (2017). With a Little Help from My Friends: The Irish Radio Industry's Strategic Appropriation of Facebook for Commercial Growth. In V. Benson, R. Tuninga, & G. Saridakis (Eds.), *Analyzing the Strategic Role of Social Networking in Firm Growth and Productivity* (pp. 157–171). Hershey, PA: IGI Global. doi:10.4018/978-1-5225-0559-4.ch009

McPherson, M. J., & Lemon, N. (2017). The Hook, Woo, and Spin: Academics Creating Relations on Social Media. In A. Esposito (Ed.), *Research 2.0 and the Impact of Digital Technologies on Scholarly Inquiry* (pp. 167–187). Hershey, PA: IGI Global. doi:10.4018/978-1-5225-0830-4.ch009

Melro, A., & Oliveira, L. (2018). Screen Culture. In M. Khosrow-Pour, D.B.A. (Ed.), Encyclopedia of Information Science and Technology, Fourth Edition (pp. 4255-4266). Hershey, PA: IGI Global. doi:10.4018/978-1-5225-2255-3.ch369

Meral, K. Z. (2021). Social Media Ethics and Children in the Digital Era: Social Media Risks and Precautions. In M. Taskiran & F. Pinarbaşi (Eds.), *Multidisciplinary Approaches to Ethics in the Digital Era* (pp. 166–182). IGI Global. https://doi.org/10.4018/978-1-7998-4117-3.ch011

Meral, Y., & Özbay, D. E. (2020). Electronic Trading, Electronic Advertising, and Social Media Literacy: Using Local Turkish Influencers in Social Media for International Trade Products Marketing. In N. Taskiran (Ed.), *Handbook of Research on Multidisciplinary Approaches to Literacy in the Digital Age* (pp. 224–261). IGI Global. https://doi.org/10.4018/978-1-7998-1534-1.ch012

Mhiripiri, N. A., & Chikakano, J. (2017). Criminal Defamation, the Criminalisation of Expression, Media and Information Dissemination in the Digital Age: A Legal and Ethical Perspective. In N. Mhiripiri & T. Chari (Eds.), *Media Law, Ethics, and Policy in the Digital Age* (pp. 1–24). Hershey, PA: IGI Global. doi:10.4018/978-1-5225-2095-5.ch001

Miliopoulou, G., & Cossiavelou, V. (2019). Brand Management and Media Gatekeeping: Exploring the Professionals' Practices and Perspectives in the Social Media. In N. Meghanathan (Ed.), *Strategic Innovations and Interdisciplinary Perspectives in Telecommunications and Networking* (pp. 56–82). IGI Global. https://doi.org/10.4018/978-1-5225-8188-8.ch004

Miranda, S. L., & Antunes, A. C. (2021). Golden Years in Social Media World: Examining Behavior and Motivations. In P. Wamuyu (Ed.), *Analyzing Global Social Media Consumption* (pp. 261–276). IGI Global. https://doi.org/10.4018/978-1-7998-4718-2.ch014

Miron, E., Palmor, A., Ravid, G., Sharon, A., Tikotsky, A., & Zirkel, Y. (2017). Principles and Good Practices for Using Wikis within Organizations. In R. Chugh (Ed.), *Harnessing Social Media as a Knowledge Management Tool* (pp. 143–176). Hershey, PA: IGI Global. doi:10.4018/978-1-5225-0495-5.ch008

Moeller, C. L. (2018). Sharing Your Personal Medical Experience Online: Is It an Irresponsible Act or Patient Empowerment? In S. Sekalala & B. Niezgoda (Eds.), *Global Perspectives on Health Communication in the Age of Social Media* (pp. 185–209). Hershey, PA: IGI Global. doi:10.4018/978-1-5225-3716-8.ch007

Mosanako, S. (2017). Broadcasting Policy in Botswana: The Case of Botswana Television. In O. Nelson, B. Ojebuyi, & A. Salawu (Eds.), *Impacts of the Media on African Socio-Economic Development* (pp. 217–230). Hershey, PA: IGI Global. doi:10.4018/978-1-5225-1859-4.ch014

Mukherjee Das, M. (2020). Harnessing the "Crowd" and the Rise of "Prosumers" in Filmmaking in India. In S. Biswal, K. Kusuma, & S. Mohanty (Eds.), *Handbook of Research on Social and Cultural Dynamics in Indian Cinema* (pp. 350–359). IGI Global. https://doi.org/10.4018/978-1-7998-3511-0.ch029

Musemburi, D., & Nhendo, C. (2019). Media Information Literacy: The Answer to 21st Century Inclusive Information and Knowledge-Based Society Challenges. In C. Chisita & A. Rusero (Eds.), *Exploring the Relationship Between Media, Libraries, and Archives* (pp. 102–135). IGI Global. https://doi.org/10.4018/978-1-5225-5840-8.ch007

Noor, R. (2017). Citizen Journalism: News Gathering by Amateurs. In M. Adria & Y. Mao (Eds.), *Handbook of Research on Citizen Engagement and Public Participation in the Era of New Media* (pp. 194–229). Hershey, PA: IGI Global. doi:10.4018/978-1-5225-1081-9.ch012

Obermayer, N., Csepregi, A., & Kővári, E. (2017). Knowledge Sharing Relation to Competence, Emotional Intelligence, and Social Media Regarding Generations. In A. Bencsik (Ed.), *Knowledge Management Initiatives and Strategies in Small and Medium Enterprises* (pp. 269–290). Hershey, PA: IGI Global. doi:10.4018/978-1-5225-1642-2.ch013

Obermayer, N., Gaál, Z., Szabó, L., & Csepregi, A. (2017). Leveraging Knowledge Sharing over Social Media Tools. In R. Chugh (Ed.), *Harnessing Social Media as a Knowledge Management Tool* (pp. 1–24). Hershey, PA: IGI Global. doi:10.4018/978-1-5225-0495-5.ch001

Odebiyi, S. D., & Elegbe, O. (2020). Human Rights Abuses Against Internally Displaced Persons (IDPs) in Nigeria: Investigating Media Reportage. In N. Okorie, B. Ojebuyi, & J. Macharia (Eds.), *Handbook of Research on the Global Impact of Media on Migration Issues* (pp. 180–200). IGI Global. https://doi.org/10.4018/978-1-7998-0210-5.ch011

Okoroafor, O. E. (2018). New Media Technology and Development Journalism in Nigeria. In A. Salawu & T. Owolabi (Eds.), *Exploring Journalism Practice and Perception in Developing Countries* (pp. 105–125). Hershey, PA: IGI Global. doi:10.4018/978-1-5225-3376-4.ch006

Okpara, S. N. (2020). Child Protection and Development in Nigeria: Towards a More Functional Media Intervention. In O. Oyero (Ed.), *Media and Its Role in Protecting the Rights of Children in Africa* (pp. 57–79). IGI Global. https://doi.org/10.4018/978-1-7998-0329-4.ch004

Olaleye, S. A., Sanusi, I. T., & Ukpabi, D. C. (2018). Assessment of Mobile Money Enablers in Nigeria. In F. Mtenzi, G. Oreku, D. Lupiana, & J. Yonazi (Eds.), *Mobile Technologies and Socio-Economic Development in Emerging Nations* (pp. 129–155). Hershey, PA: IGI Global. doi:10.4018/978-1-5225-4029-8.ch007

Pacchiega, C. (2017). An Informal Methodology for Teaching Through Virtual Worlds: Using Internet Tools and Virtual Worlds in a Coordinated Pattern to Teach Various Subjects. In G. Panconesi & M. Guida (Eds.), *Handbook of Research on Collaborative Teaching Practice in Virtual Learning Environments* (pp. 163–180). Hershey, PA: IGI Global. doi:10.4018/978-1-5225-2426-7.ch009

Pant, L. D. (2021). Gender Mainstreaming in the Media: The Issue of Professional and Workplace Safety of Women Journalists in Nepal. In S. Jamil, B. Çoban, B. Ataman, & G. Appiah-Adjei (Eds.), *Handbook of Research on Discrimination, Gender Disparity, and Safety Risks in Journalism* (pp. 194–210). IGI Global. https://doi.org/10.4018/978-1-7998-6686-2.ch011

Pase, A. F., Goss, B. M., & Tietzmann, R. (2018). A Matter of Time: Transmedia Journalism Challenges. In R. Gambarato & G. Alzamora (Eds.), *Exploring Transmedia Journalism in the Digital Age* (pp. 49–66). Hershey, PA: IGI Global. doi:10.4018/978-1-5225-3781-6.ch004

Patkin, T. T. (2017). Social Media and Knowledge Management in a Crisis Context: Barriers and Opportunities. In R. Chugh (Ed.), *Harnessing Social Media as a Knowledge Management Tool* (pp. 125–142). Hershey, PA: IGI Global. doi:10.4018/978-1-5225-0495-5.ch007

Pavlíček, A. (2017). Social Media and Creativity: How to Engage Users and Tourists. In A. Kiráľová (Ed.), *Driving Tourism through Creative Destinations and Activities* (pp. 181–202). Hershey, PA: IGI Global. doi:10.4018/978-1-5225-2016-0.ch009

Pérez-Gómez, M. Á. (2020). Augmented Reality and Franchising: The Evolution of Media Mix Through Invizimals. In V. Hernández-Santaolalla & M. Barrientos-Bueno (Eds.), *Handbook of Research on Transmedia Storytelling, Audience Engagement, and Business Strategies* (pp. 90–102). IGI Global. https://doi.org/10.4018/978-1-7998-3119-8.ch007

Phiri, S., & Mokorosi, L. (2020). Of Elephants and Men: Understanding Gender-Based Hate Speech in Zambia's Social Media Platforms. In J. Kurebwa (Ed.), *Understanding Gender in the African Context* (pp. 105–125). IGI Global. https://doi.org/10.4018/978-1-7998-2815-0.ch006

Pillai, A. P. (2019). Nuances of Media Planning in New Media Age. In P. Mishra & S. Dham (Eds.), *Application of Gaming in New Media Marketing* (pp. 151–170). IGI Global. https://doi.org/10.4018/978-1-5225-6064-7.ch009

Pillay, K., & Maharaj, M. (2017). The Business of Advocacy: A Case Study of Greenpeace. In V. Benson, R. Tuninga, & G. Saridakis (Eds.), *Analyzing the Strategic Role of Social Networking in Firm Growth and Productivity* (pp. 59–75). Hershey, PA: IGI Global. doi:10.4018/978-1-5225-0559-4.ch004

Piven, I. P., & Breazeale, M. (2017). Desperately Seeking Customer Engagement: The Five-Sources Model of Brand Value on Social Media. In V. Benson, R. Tuninga, & G. Saridakis (Eds.), *Analyzing the Strategic Role of Social Networking in Firm Growth and Productivity* (pp. 283–313). Hershey, PA: IGI Global. doi:10.4018/978-1-5225-0559-4.ch016

Pokharel, R. (2017). New Media and Technology: How Do They Change the Notions of the Rhetorical Situations? In B. Gurung & M. Limbu (Eds.), *Integration of Cloud Technologies in Digitally Networked Classrooms and Learning Communities* (pp. 120–148). Hershey, PA: IGI Global. doi:10.4018/978-1-5225-1650-7.ch008

Porlezza, C., Benecchi, E., & Colapinto, C. (2018). The Transmedia Revitalization of Investigative Journalism: Opportunities and Challenges of the Serial Podcast. In R. Gambarato & G. Alzamora (Eds.), *Exploring Transmedia Journalism in the Digital Age* (pp. 183–201). Hershey, PA: IGI Global. doi:10.4018/978-1-5225-3781-6.ch011

Ramluckan, T., Ally, S. E., & van Niekerk, B. (2017). Twitter Use in Student Protests: The Case of South Africa's #FeesMustFall Campaign. In M. Korstanje (Ed.), *Threat Mitigation and Detection of Cyber Warfare and Terrorism Activities* (pp. 220–253). Hershey, PA: IGI Global. doi:10.4018/978-1-5225-1938-6.ch010

Rao, N. R. (2017). Social Media: An Enabler for Governance. In N. Rao (Ed.), *Social Media Listening and Monitoring for Business Applications* (pp. 151–164). Hershey, PA: IGI Global. doi:10.4018/978-1-5225-0846-5.ch008

Redi, F. (2017). Enhancing Coopetition Among Small Tourism Destinations by Creativity. In A. Kiráľová (Ed.), *Driving Tourism through Creative Destinations and Activities* (pp. 223–244). Hershey, PA: IGI Global. doi:10.4018/978-1-5225-2016-0.ch011

Resuloğlu, F., & Yılmaz, R. (2017). A Model for Interactive Advertising Narration. In R. Yılmaz (Ed.), *Narrative Advertising Models and Conceptualization in the Digital Age* (pp. 1–20). Hershey, PA: IGI Global. doi:10.4018/978-1-5225-2373-4.ch001

Richards, M. B. (2022). Media and Parental Communication: Effects on Millennials' Value Formation. In S. Malik, R. Bansal, & A. Tyagi (Eds.), *Impact and Role of Digital Technologies in Adolescent Lives* (pp. 64–82). IGI Global. https://doi.org/10.4018/978-1-7998-8318-0.ch006

Robinson, W. R. (2021). The Intellectual Soul Food Lunch Buffet: The Classroom to Student Media Entrepreneurship. In L. Byrd (Ed.), *Cultivating Entrepreneurial Changemakers Through Digital Media Education* (pp. 108–121). IGI Global. https://doi.org/10.4018/978-1-7998-5808-9.ch007

Ross, D. B., Eleno-Orama, M., & Salah, E. V. (2018). The Aging and Technological Society: Learning Our Way Through the Decades. In V. Bryan, A. Musgrove, & J. Powers (Eds.), *Handbook of Research on Human Development in the Digital Age* (pp. 205–234). Hershey, PA: IGI Global. doi:10.4018/978-1-5225-2838-8.ch010

Rusko, R., & Merenheimo, P. (2017). Co-Creating the Christmas Story: Digitalizing as a Shared Resource for a Shared Brand. In I. Oncioiu (Ed.), *Driving Innovation and Business Success in the Digital Economy* (pp. 137–157). Hershey, PA: IGI Global. doi:10.4018/978-1-5225-1779-5.ch010

Sabao, C., & Chikara, T. O. (2018). Social Media as Alternative Public Sphere for Citizen Participation and Protest in National Politics in Zimbabwe: The Case of #thisflag. In F. Endong (Ed.), *Exploring the Role of Social Media in Transnational Advocacy* (pp. 17–35). Hershey, PA: IGI Global. doi:10.4018/978-1-5225-2854-8.ch002

Saçak, B. (2019). Media Literacy in a Digital Age: Multimodal Social Semiotics and Reading Media. In M. Yildiz, M. Fazal, M. Ahn, R. Feirsen, & S. Ozdemir (Eds.), *Handbook of Research on Media Literacy Research and Applications Across Disciplines* (pp. 13–26). IGI Global. https://doi.org/10.4018/978-1-5225-9261-7.ch002

Samarthya-Howard, A., & Rogers, D. (2018). Scaling Mobile Technologies to Maximize Reach and Impact: Partnering With Mobile Network Operators and Governments. In S. Takavarasha Jr & C. Adams (Eds.), *Affordability Issues Surrounding the Use of ICT for Development and Poverty Reduction* (pp. 193–211). Hershey, PA: IGI Global. doi:10.4018/978-1-5225-3179-1.ch009

Sandoval-Almazan, R. (2017). Political Messaging in Digital Spaces: The Case of Twitter in Mexico's Presidential Campaign. In Y. Ibrahim (Ed.), *Politics, Protest, and Empowerment in Digital Spaces* (pp. 72–90). Hershey, PA: IGI Global. doi:10.4018/978-1-5225-1862-4.ch005

Schultz, C. D., & Dellnitz, A. (2018). Attribution Modeling in Online Advertising. In K. Yang (Ed.), *Multi-Platform Advertising Strategies in the Global Marketplace* (pp. 226–249). Hershey, PA: IGI Global. doi:10.4018/978-1-5225-3114-2.ch009

Schultz, C. D., & Holsing, C. (2018). Differences Across Device Usage in Search Engine Advertising. In K. Yang (Ed.), *Multi-Platform Advertising Strategies in the Global Marketplace* (pp. 250–279). Hershey, PA: IGI Global. doi:10.4018/978-1-5225-3114-2.ch010

Seçkin, G. (2020). The Integration of the Media With the Power in Turkey (2002-2019): Native, National Media Conception. In S. Karlidag & S. Bulut (Eds.), *Handbook of Research on the Political Economy of Communications and Media* (pp. 206–226). IGI Global. https://doi.org/10.4018/978-1-7998-3270-6.ch011

Senadheera, V., Warren, M., Leitch, S., & Pye, G. (2017). Facebook Content Analysis: A Study into Australian Banks' Social Media Community Engagement. In S. Hai-Jew (Ed.), *Social Media Data Extraction and Content Analysis* (pp. 412–432). Hershey, PA: IGI Global. doi:10.4018/978-1-5225-0648-5.ch013

Sharma, A. R. (2018). Promoting Global Competencies in India: Media and Information Literacy as Stepping Stone. In M. Yildiz, S. Funk, & B. De Abreu (Eds.), *Promoting Global Competencies Through Media Literacy* (pp. 160–174). Hershey, PA: IGI Global. doi:10.4018/978-1-5225-3082-4.ch010

Sharma, D., & Bhattacharya, S. (2022). Complexity of Digital Media Crowning the Mental Health of Adolescents. In S. Malik, R. Bansal, & A. Tyagi (Eds.), *Impact and Role of Digital Technologies in Adolescent Lives* (pp. 100–117). IGI Global. https://doi.org/10.4018/978-1-7998-8318-0.ch008

Sillah, A. (2017). Nonprofit Organizations and Social Media Use: An Analysis of Nonprofit Organizations' Effective Use of Social Media Tools. In M. Brown Sr., (Ed.), *Social Media Performance Evaluation and Success Measurements* (pp. 180–195). Hershey, PA: IGI Global. doi:10.4018/978-1-5225-1963-8.ch009

Silva, H., & Simão, E. (2019). Thinking Art in the Technological World: An Approach to Digital Media Art Creation. In E. Simão & C. Soares (Eds.), *Trends, Experiences, and Perspectives in Immersive Multimedia and Augmented Reality* (pp. 102–121). IGI Global. https://doi.org/10.4018/978-1-5225-5696-1.ch005

Škorić, M. (2017). Adaptation of Winlink 2000 Emergency Amateur Radio Email Network to a VHF Packet Radio Infrastructure. In A. El Oualkadi & J. Zbitou (Eds.), *Handbook of Research on Advanced Trends in Microwave and Communication Engineering* (pp. 498–528). Hershey, PA: IGI Global. doi:10.4018/978-1-5225-0773-4.ch016

Soares, C., & Simão, E. (2020). Software-Based Media Art: From the Artistic Exhibition to the Conservation Models. In C. Soares & E. Simão (Eds.), *Multidisciplinary Perspectives on New Media Art* (pp. 47–63). IGI Global. https://doi.org/10.4018/978-1-7998-3669-8.ch003

Sonnenberg, C. (2020). Mobile Media Usability: Evaluation of Methods for Adaptation and User Engagement. *Journal of Media Management and Entrepreneurship*, *2*(1), 86–107. https://doi.org/10.4018/JMME.2020010106

Sood, T. (2017). Services Marketing: A Sector of the Current Millennium. In T. Sood (Ed.), *Strategic Marketing Management and Tactics in the Service Industry* (pp. 15–42). Hershey, PA: IGI Global. doi:10.4018/978-1-5225-2475-5.ch002

Sudarsanam, S. K. (2017). Social Media Metrics. In N. Rao (Ed.), *Social Media Listening and Monitoring for Business Applications* (pp. 131–149). Hershey, PA: IGI Global. doi:10.4018/978-1-5225-0846-5.ch007

Swiatek, L. (2017). Accessing the Finest Minds: Insights into Creativity from Esteemed Media Professionals. In N. Silton (Ed.), *Exploring the Benefits of Creativity in Education, Media, and the Arts* (pp. 240–263). Hershey, PA: IGI Global. doi:10.4018/978-1-5225-0504-4.ch012

Teurlings, J. (2017). What Critical Media Studies Should Not Take from Actor-Network Theory. In M. Spöhrer & B. Ochsner (Eds.), *Applying the Actor-Network Theory in Media Studies* (pp. 66–78). Hershey, PA: IGI Global. doi:10.4018/978-1-5225-0616-4.ch005

Tilwankar, V., Rai, S., & Bajpai, S. P. (2019). Role of Social Media in Environment Awareness: Social Media and Environment. In S. Narula, S. Rai, & A. Sharma (Eds.), *Environmental Awareness and the Role of Social Media* (pp. 117–139). IGI Global. https://doi.org/10.4018/978-1-5225-5291-8.ch006

Tokbaeva, D. (2019). Media Entrepreneurs and Market Dynamics: Case of Russian Media Markets. *Journal of Media Management and Entrepreneurship*, *1*(1), 40–56. https://doi.org/10.4018/JMME.2019010103

Tomé, V. (2018). Assessing Media Literacy in Teacher Education. In M. Yildiz, S. Funk, & B. De Abreu (Eds.), *Promoting Global Competencies Through Media Literacy* (pp. 1–19). Hershey, PA: IGI Global. doi:10.4018/978-1-5225-3082-4.ch001

Topçu, Ç. (2022). Social Media and the Knowledge Gap: Research on the Appearance of COVID-19 in Turkey and the Knowledge Level of Users. In H. Aker & M. Aiken (Eds.), *Handbook of Research on Cyberchondria, Health Literacy, and the Role of Media in Society's Perception of Medical Information* (pp. 344–361). IGI Global. https://doi.org/10.4018/978-1-7998-8630-3.ch019

Toscano, J. P. (2017). Social Media and Public Participation: Opportunities, Barriers, and a New Framework. In M. Adria & Y. Mao (Eds.), *Handbook of Research on Citizen Engagement and Public Participation in the Era of New Media* (pp. 73–89). Hershey, PA: IGI Global. doi:10.4018/978-1-5225-1081-9.ch005

Trauth, E. (2017). Creating Meaning for Millennials: Bakhtin, Rosenblatt, and the Use of Social Media in the Composition Classroom. In K. Bryant (Ed.), *Engaging 21st Century Writers with Social Media* (pp. 151–162). Hershey, PA: IGI Global. doi:10.4018/978-1-5225-0562-4.ch009

Trucks, E. (2019). Making Social Media More Social: A Literature Review of Academic Libraries' Engagement and Connections Through Social Media Platforms. In J. Joe & E. Knight (Eds.), *Social Media for Communication and Instruction in Academic Libraries* (pp. 1–16). IGI Global. https://doi.org/10.4018/978-1-5225-8097-3.ch001

Udenze, S. (2021). Social Media and Nigeria's Politics. In S. Aririguzoh (Ed.), *Global Perspectives on the Impact of Mass Media on Electoral Processes* (pp. 83–96). IGI Global. https://doi.org/10.4018/978-1-7998-4820-2.ch005

Uprety, S. (2018). Print Media's Role in Securitization: National Security and Diplomacy Discourses in Nepal. In S. Bute (Ed.), *Media Diplomacy and Its Evolving Role in the Current Geopolitical Climate* (pp. 56–82). Hershey, PA: IGI Global. doi:10.4018/978-1-5225-3859-2.ch004

Uprety, S., & Chand, O. B. (2021). Trump's Declaration of the Global Gag Rule: Understanding Socio-Political Discourses Through Media. In E. Hancı-Azizoglu & M. Alawdat (Eds.), *Rhetoric and Sociolinguistics in Times of Global Crisis* (pp. 277–294). IGI Global. https://doi.org/10.4018/978-1-7998-6732-6.ch015

van der Vyver, A. G. (2018). A Model for Economic Development With Telecentres and the Social Media: Overcoming Affordability Constraints. In S. Takavarasha Jr & C. Adams (Eds.), *Affordability Issues Surrounding the Use of ICT for Development and Poverty Reduction* (pp. 112–140). Hershey, PA: IGI Global. doi:10.4018/978-1-5225-3179-1.ch006 van Niekerk, B. (2018). Social Media Activism From an Information Warfare and Security Perspective. In F. Endong (Ed.), *Exploring the Role of Social Media in Transnational Advocacy* (pp. 1–16). Hershey, PA: IGI Global. doi:10.4018/978-1-5225-2854-8.ch001

Varnali, K., & Gorgulu, V. (2017). Determinants of Brand Recall in Social Networking Sites. In W. Al-Rabayah, R. Khasawneh, R. Abu-shamaa, & I. Alsmadi (Eds.), *Strategic Uses of Social Media for Improved Customer Retention* (pp. 124–153). Hershey, PA: IGI Global. doi:10.4018/978-1-5225-1686-6.ch007

Varty, C. T., O'Neill, T. A., & Hambley, L. A. (2017). Leading Anywhere Workers: A Scientific and Practical Framework. In Y. Blount & M. Gloet (Eds.), *Anywhere Working and the New Era of Telecommuting* (pp. 47–88). Hershey, PA: IGI Global. doi:10.4018/978-1-5225-2328-4.ch003

Velikovsky, J. T. (2018). The Holon/Parton Structure of the Meme, or The Unit of Culture. In M. Khosrow-Pour, D.B.A. (Ed.), Encyclopedia of Information Science and Technology, Fourth Edition (pp. 4666-4678). Hershey, PA: IGI Global. https://doi.org/ doi:10.4018/978-1-5225-2255-3.ch405

Venkatesh, R., & Jayasingh, S. (2017). Transformation of Business through Social Media. In N. Rao (Ed.), *Social Media Listening and Monitoring for Business Applications* (pp. 1–17). Hershey, PA: IGI Global. doi:10.4018/978-1-5225-0846-5.ch001

Vijayakumar, D. S., M., S., Thangaraju, J., & V., S. (2021). Social Media Content Analysis: Machine Learning. In V. Sathiyamoorthi, & A. Elci (Eds.), *Challenges and Applications of Data Analytics in Social Perspectives* (pp. 156-174). IGI Global. https://doi.org/10.4018/978-1-7998-2566-1.ch009

Virkar, S. (2017). Trolls Just Want to Have Fun: Electronic Aggression within the Context of E-Participation and Other Online Political Behaviour in the United Kingdom. In M. Korstanje (Ed.), *Threat Mitigation and Detection of Cyber Warfare and Terrorism Activities* (pp. 111–162). Hershey, PA: IGI Global. doi:10.4018/978-1-5225-1938-6.ch006

Wakabi, W. (2017). When Citizens in Authoritarian States Use Facebook for Social Ties but Not Political Participation. In Y. Ibrahim (Ed.), *Politics, Protest, and Empowerment in Digital Spaces* (pp. 192–214). Hershey, PA: IGI Global. doi:10.4018/978-1-5225-1862-4.ch012

Wamuyu, P. K. (2021). Social Media Consumption Among Kenyans: Trends and Practices. In P. Wamuyu (Ed.), *Analyzing Global Social Media Consumption* (pp. 88–120). IGI Global. https://doi.org/10.4018/978-1-7998-4718-2.ch006

Wright, K. (2018). "Show Me What You Are Saying": Visual Literacy in the Composition Classroom. In A. August (Ed.), *Visual Imagery, Metadata, and Multimodal Literacies Across the Curriculum* (pp. 24–49). Hershey, PA: IGI Global. doi:10.4018/978-1-5225-2808-1.ch002

Wright, M. F. (2020). Cyberbullying: Negative Interaction Through Social Media. In M. Desjarlais (Ed.), *The Psychology and Dynamics Behind Social Media Interactions* (pp. 107–135). IGI Global. https://doi.org/10.4018/978-1-5225-9412-3.ch005

Yang, K. C. (2018). Understanding How Mexican and U.S. Consumers Decide to Use Mobile Social Media: A Cross-National Qualitative Study. In K. Yang (Ed.), *Multi-Platform Advertising Strategies in the Global Marketplace* (pp. 168–198). Hershey, PA: IGI Global. doi:10.4018/978-1-5225-3114-2.ch007

Yarchi, M., Wolfsfeld, G., Samuel-Azran, T., & Segev, E. (2017). Invest, Engage, and Win: Online Campaigns and Their Outcomes in an Israeli Election. In M. Brown Sr., (Ed.), *Social Media Performance Evaluation and Success Measurements* (pp. 225–248). Hershey, PA: IGI Global. doi:10.4018/978-1-5225-1963-8.ch011

Yeboah-Banin, A. A., & Amoakohene, M. I. (2018). The Dark Side of Multi-Platform Advertising in an Emerging Economy Context. In K. Yang (Ed.), *Multi-Platform Advertising Strategies in the Global Marketplace* (pp. 30–53). Hershey, PA: IGI Global. doi:10.4018/978-1-5225-3114-2.ch002

Yılmaz, R., Çakır, A., & Resuloğlu, F. (2017). Historical Transformation of the Advertising Narration in Turkey: From Stereotype to Digital Media. In R. Yılmaz (Ed.), *Narrative Advertising Models and Conceptualization in the Digital Age* (pp. 133–152). Hershey, PA: IGI Global. doi:10.4018/978-1-5225-2373-4.ch008

Yusuf, S., Hassan, M. S., & Ibrahim, A. M. (2018). Cyberbullying Among Malaysian Children Based on Research Evidence. In M. Khosrow-Pour, D.B.A. (Ed.), Encyclopedia of Information Science and Technology, Fourth Edition (pp. 1704-1722). Hershey, PA: IGI Global. doi:10.4018/978-1-5225-2255-3.ch149

Zbinden, B. (2019). Restricted Communication: Social Relationships and the Media Use of Prisoners. In L. Oliveira & D. Graça (Eds.), *Infocommunication Skills as a Rehabilitation and Social Reintegration Tool for Inmates* (pp. 238–267). IGI Global. doi:10.4018/978-1-5225-5975-7.ch011

Zhou, M., Matsika, C., Zhou, T. G., & Chawarura, W. I. (2022). Harnessing Social Media to Improve Educational Performance of Adolescent Freshmen in Universities. In S. Malik, R. Bansal, & A. Tyagi (Eds.), *Impact and Role of Digital Technologies in Adolescent Lives* (pp. 51–63). IGI Global. https://doi.org/10.4018/978-1-7998-8318-0.ch005

# About the Contributors

**T. S. Pradeep Kumar** is a Professor from Vellore Institute of Technology Chennai, holds a Bachelors (Electrical & Electronics) and Master's degree in Embedded System Technologies and PhD in the area of Power modelling of Sensors for Internet of Things. He published more than 25 peer reviewed journals and been session chairs for many conferences. He is interested towards open source computing and research. He had deployed and maintain the MOODLE LMS and BigBlueButton Virtual Classroom Software in the campus that handles the requirements of more than 15000 users of an higher educational institution. His research interests includes Internet of Things, Sensor Networks, Wireless Adhoc Networks, Embedded Systems and E Learning.

**M. Alamelu** working as Associate Professor and leading Head of Information Technology, Kumaraguru College of Technology, Coimbatore. She has 15 years of experience in teaching and research and completed her Doctorate in the field Service Oriented Architecture and published several articles in the National/International Journals and conferences. Her research area includes Service Oriented Architecture, Web Services, Data Mining and Machine Learning. She has organized several national workshops, FDPs, seminars and social outreach programs. Specifically, the social outreach programs like "IT Technical Training program", "Computing technology", "Recent trends in Internet of Things (IoT)", "CODE INFO", "Computing technical awareness programme", STTP for "Enlighten the programming Basics" was conducted for the rural school students to enhance their computer-based knowledge. She was also part of IEEE funded project "Smart Agriculture for sustainable food production" sponsored by IEEE foundation. The project focuses on the enhancement of smart agriculture, to expand the crop yielding technologies with respect to the soil conditions and reach out the smart agricultural Information Technology technique awareness to farmers. And she has written book chapters, Filed patent, organized Runners club and received 1 International award, 1 National Award and 7 awards at Kumaraguru College of Technology, Coimbatore. Her special interest includes the conduction of Information Technology Technical programs for the disabled persons.

* * *

**S. Vijay Anand** is working as an assistant professor in department of Electronics and Communication Engineering, Sri Venkateswara College of Engineering, Chennai. He is having more than 17 years of teaching experience in ECE. He published more than fifteen journal papers in various journals and publishing book chapters. He published three patterns.

**Sathis Kumar B.** is a Professor from Vellore Institute of Technology Chennai, holds Master's degree in Computer Science and Engineering. His research interests include Software Engineering, Software Architecture, Gamification, Education Technology, Social Computing and Networks

**Rafik Bouaziz** is a full professor of Computer Science at the Faculty of Economics and Management of Sfax University, Tunisia. He was the president of this University during August 2014 – December 2017, and the director of its doctoral school of economy, management and computer science during December 2011 – July 2014. His PhD has dealt with temporal data management and historical record of data in Information Systems. The subject of his accreditation to supervise research was "A contribution for the control of versioning of data and schema in advanced information systems". Currently, his main research topics of interest are temporal databases, real-time databases, information systems engineering, ontologies, data warehousing and workflows. Between 1979 and 1986, he was a consulting Engineer in the organization and computer science and a head of the department of computer science at CEGOS-TUNISIA.

**Zouhaier Brahmia** is currently an Associate Professor of Computer Science in the Department of Computer Science at the Faculty of Economics and Management of the University of Sfax, Tunisia. He is a member of the Multimedia, InfoRmation systems, and Advanced Computing Laboratory (MIRACL). His scientific interests include temporal databases, schema versioning, and temporal, evolution and versioning aspects in emerging databases (XML, and NoSQL), Big Data, ontologies, World Wide Web, and Semantic Web. He received an MSc degree in Computer Science, in July 2005, and a PhD in Computer Science, in December 2011, from the Faculty of Economics and Management of the University of Sfax.

**Sankar Ram C.** is working as Assistant Professor at Anna University Tiruchirappalli. Completed two AICTE projects in 2019 and 2020 and having 20 publications. Got best mentor award from Government of India for IICDC (AICTE, DST and Texas Instruments).

**Fabio Grandi** is currently an Associate Professor in the School of Engineering of the University of Bologna, Italy. Since 1989 he has worked at the CSITE center of the Italian National Research Council (CNR) in Bologna in the field of neural networks and temporal databases, initially supported by a CNR fellowship. In 1993 and 1994 he was an Adjunct Professor at the Universities of Ferrara, Italy, and Bologna. In the University of Bologna, he was with the Dept. of Electronics, Computers and Systems from 1994 to 1998 as a Research Associate and as Associate Professor from 1998 to 2012, when he joined the Dept. of Computer Science and Engineering. His scientific interests include temporal, evolution and versioning aspects in data management, WWW and Semantic Web, knowledge representation, storage structures and access cost models. He received a Laurea degree cum Laude in Electronics Engineering and a PhD in Electronics Engineering and Computer Science from the University of Bologna.

**Sushmitha J.** is currently working as an Assistant Professor in Rohini College of Engineering, Kanniyakumari. She completed her B.Em in Ponjesly College of Engineering and M.Em in Government College of engineering and technology, Tirunelveli. She is a gold medalist in PG. (Communication systems). Her active interests include wireless adhoc networks, MANETS and networking technology. She is pursuing her PhD in Anna University.

**Pabitra Mohan Khilar** received his Ph.D. in Computer Science and Engineering in 2009 from Indian Institute of Technology (IIT), Kharagpur, India. He received his M.Tech degree in 1999 from the Department of Computer Science and Engineering, National Institute of Technology (NIT), Rourkela, India. He received his B.Tech degree in 1990 from Department of Computer Science and Engineering, University of Mysore, India. He is currently an Associate Professor in the Department of Computer Science and Engineering, National Institute of Technology (NIT), Rourkela, India. His research interests include Parallel and Distributed Computing, Fault Tolerant Systems, Wireless Networks, Sensors, and VLSI Design. He has a teaching experience of more than 20 years. He has awarded 7 doctoral degrees. He has authored more than 150 journals and conference papers. He has published papers in journals like Wireless Networks, Wireless Personal Communications, IJCS Wiley, Ad Hoc Networks, Computer and Electrical Engineering, Swarm and Evolutionary Computation, IET Networks, IET Wireless Sensor Systems, and IEEE Surveys and Tutorials. He received the IET Premium Award-2016 from IET Networks for the best journal paper in the last two years. He is a Life Member in MIEEE, MIE, MCSI, MISTE, MOEC, MOITS, MIETE, and MISCA.

**Nallarasu Krishnan** is presently with Tagore Engineering College, Chennai, India. He is Assistant Professor in the department of Computer Science and Engineering. He is now a part-time research scholar registered with Anna University, Chennai, India. Currently he is doing research in the area of Wireless Sensor Networks.

**Divya Lanka** is currently working as assistant professor in Department of CSE, SRKR Engineering College. She is pursuing Ph.D. In Pondicherry Central University. Her research areas of interest are Internet of Things and Software Defined Networks.

**Dhirendra Pandey** is an Assistant Professor of Information Technology at Babasaheb Bhimrao Ambedkar University, India. He received his doctorate in Computer Science from Devi Ahilya Vishwavidyalaya, Indore, India. His areas of specializations are Data Mining, Software Engineering, and Requirement Engineering. He has more than 8 years of teaching experience and research experience. He is the reviewer of many national and international journals. In addition, He has also published many research papers as an author and co-author of more than 25 papers in national and international journals. Dr. Dhirendra Pandey also working as Principal Investigator of Research Project funded by CST, UP, India.

**Vetrivelan Pandu**, PhD, is an Associate Professor Senior and Head of the Department (HoD) for Bachelor of Technology (Electronics and Communication Engineering) in School of Electronics Engineering at Vellore Institute of Technology (VIT), Chennai, India. He has completed Bachelor of Engineering from the University of Madras, Chennai and both Master of Engineering in Embedded Systems Technologies and Doctor of Philosophy in Information and Communication Engineering from Anna University, Chennai. He has 17.3 years of teaching experience altogether in CSE and ECE Departments in both private Engineering Colleges in Chennai (affiliated to Anna University, Chennai) and Private Engineering University in Chennai respectively. He has authored 3 book chapters and one proceeding in lecture notes published by reputed springer publisher, and has authored 25+ Scopus indexed Journal papers and few other papers published in reputed international conferences. He has served as member in Board of Studies, doctoral committee, doctoral thesis Examiner, doctoral oral Examiner in both private and government Universities. He has also serves as reviewer for reputed International Journals and International Conferences. His research interests include Wireless Networks, Adhoc and Sensor Networks, VANETs, Embedded Systems and Internet of Things (IoT) with Machine Learning.

**Sudarson Rama Perumal** is working as a Associate Professor in the department of electronics and communication engineering in Rohini College of Engineering,

Kanniyakumari. He completed his PhD in Anna University, Chennai. His active research interests include Wireless adhoc networks, image processing, bio medical imaging and networking.

**Uma R.** is currently working as Associate Professor in the Department of Computer Science and Engineering at Vardhaman College of Engineering, Hyderabad, Telangana, India. She completed her B.Tech in Information Technology from Anna University and M.Tech in Computer and Information Technology from Manonmaniam Sundaranar University. She completed her Ph.D in Information and Communication Engineering from Anna University. She has teaching experience of more than 13 years. She has several publications in reputed Journals and Conferences. She is a member of IEEE, ACM and SDIWC. Her research interest includes Wireless networks, IoT, Neural networks and Machine Learning.

**Biswa Ranjan Senapati** is pursuing his Ph.D. in the Department of Computer Science and Engineering, National Institute of Technology (NIT), Rourkela, India. He received his M.Tech. degree in 2011 from the Department of Computer Science and Engineering, CET Bhubaneswar, India. He has received his B.Tech. degree in 2009 from Department of Computer Science and Engineering, ITER Bhubaneswar, India. His research interests include Wireless Sensor Networks, Vehicular Ad-Hoc Networks, and IoT.

**Sipra Swain** is pursuing her PhD degree in Computer Science and Engineering department of National Institute of Technology Rourkela, India. She has received her M.Tech degree in Computer Science and Engineering department of National Institute of Technology Durgapur, India in 2019. She has completed her B.Tech degree in Computer Science and Engineering department of Biju Patnaik University of Technology, India in 2016. Her research area of interest includes UAV, wireless sensor network, distributed computing, and swarm robotics.

**Muthumanikandan V.**, B.E., M.E., Ph.D., is working as a Senior Assistant Professor in the School of Computing Science and Engineering, at Vellore Institute of Technology, Chennai, India. He received his B.E and M.E degree in Computer Science and Engineering discipline. He received his Ph.D degree in Computer Science and Engineering from Anna University. His areas of interests include Networking, Software Defined Networking and Network Function Virtualization.

**Vanitha V.** has completed B.E.(CSE) at Institute of Road and Transport Technology, Erode in 1994, M.E.(CSE) at Kumaraguru College of Technology in 2001 and Ph.D in ICE at Anna University Chennai, in 2012. She is presently working as

Professor in the Department of Information Technology. She has 20 years of teaching experience in KCT and 2 years of industry experience. She is a life member of ISTE. She has attended more than 40 quality improvement programmes so far. She has organized DRDO sponsored workshop on "Wireless Sensor Networks". Now-a-days she is actively conducting number of workshops and 1-credit courses on "Internet of Things", "Arduino Programming" and "Raspberry Pi Programming" in KCT and in other institutions like SNS, Jeppiar Institute of Technology, Chennai, Sri Venkateshwara College of Engineering, Bangalore and Govt. Arts College Salem. Also she has delivered number of guest lectures. She has around 25 publications in National and International journals. Her research interests include computer networks, Wireless Sensor Networks and Internet of Things. She is the co-author for three books.

**Mohammed Sirajudeen Yoosuf** is an Assistant Professor with a demonstrated history of working on cloud technology and security. Skilled in AWS, python programming language, Hadoop, and Machine Learning. Strong research professional with a Ph.D. degree focused on cyber governance from Anna University, Chennai, India. My areas of interest are distributed systems, IoT, Drone Technology, cyber security, AI, and ML.

# Index

Printed in the United States
by Baker & Taylor Publisher Services